Projective Geometry

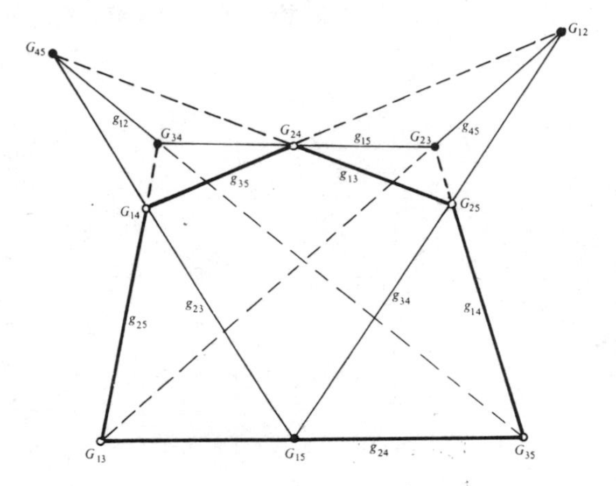

Two mutually inscribed pentagons

Projective Geometry

SECOND EDITION

H. S. M. Coxeter

UNIVERSITY OF TORONTO

UNIVERSITY OF TORONTO PRESS

TO RIEN

Toronto and Buffalo

Printed in Canada
ISBN 0-8020-2104-2
LC 73-86992

The first edition of this book was published by Blaisdell Publishing Company in 1964.

Preface to the First Edition

In Euclidean geometry, constructions are made with the ruler and compass. Projective geometry is simpler: its constructions require only the ruler. We consider the straight line joining two points, and the point of intersection of two lines, with the further simplification that two lines never fail to meet!

In Euclidean geometry we compare figures by measuring them. In projective geometry we never measure anything; instead, we relate one set of points to another by a *projectivity*. Chapter 1 introduces the reader to this important idea. Chapter 2 provides a logical foundation for the subject. The third and fourth chapters describe the famous theorems of Desargues and Pappus. The fifth and sixth make use of projectivities on a line and in a plane, respectively. In the next three we develop a self-contained account of von Staudt's approach to the theory of conics, made more "modern" by allowing the field to be general (though not of characteristic 2) instead of real or complex. This freedom has been exploited in Chapter 10, which deals with the simplest *finite* geometry that is rich enough to illustrate all our theorems nontrivially (for instance, Pascal's theorem concerns six points on a conic, and in $PG(2, 5)$ these are the *only* points on the conic). In Chapters 11 and 12 we return to more familiar ground, showing the connections between projective geometry, Euclidean geometry, and the popular subject of "analytic geometry."

The possibility of writing an easy book on projective geometry was foreseen as long ago as 1917, when D. N. Lehmer [**12,*** Preface, p. v] wrote:

> The subject of synthetic projective geometry is . . . destined shortly to force its way down into the secondary schools.

More recently, A. N. Whitehead [**22**, p. 133] recommended a revised curriculum beginning with Congruence, Similarity, Trigonometry, Analytic

---

* References are given on page 158.

Geometry, and then:

> In this ideal course of Geometry, the fifth stage is occupied with the elements of Projective Geometry . . .

This "fifth" stage has one notable advantage: its primitive concepts are so simple that a self-contained account can be reasonably entertaining, whereas the foundations of Euclidean geometry are inevitably tedious.

The present treatment owes much to the famous text-book of Veblen and Young [**19**], which has the same title. To encourage truly geometric habits of thought, we avoid the use of coordinates and all metrical ideas (Whitehead's first four "stages") except in Chapters 1, 11, 12, and a few of the Exercises. In particular, the only mention of *cross ratio* is in three exercises at the end of Section 12.3.

I gratefully acknowledge the help of M. W. Al-Dhahir, W. L. Edge, P. R. Halmos, S. Schuster and S. Trott, who constructively criticized the manuscript, and of H. G. Forder and C. Garner, who read the proofs. I wish also to express my thanks for permission to quote from *Science: Sense and Nonsense* by J. L. Synge (Jonathan Cape, London).

H. S. M. COXETER

Toronto, Canada
February, 1963

Preface to the Second Edition

Why should one study Pappian geometry? To this question, put by enthusiasts for ternary rings, I would reply that the classical projective plane is an easy first step. The theory of conics is beautiful in itself and provides a natural introduction to algebraic geometry.

Apart from the correction of many small errors, the changes made in this revised edition are chiefly as follows. Veblen's notation $Q(ABC, DEF)$ for a quadrangular set of six points has been replaced by the "permutation symbol" $(AD)(BE)(CF)$, which indicates more immediately that there is an involution interchanging the points on each pair of opposite sides of the quadrangle. Although most of the work is in the projective *plane*, it has seemed worth while (in Section 3.2) to show how the Desargues configuration can be derived as a section of the "complete 5-point" in space. Section 4.4 emphasizes the analogy between the configurations of Desargues and Pappus. At the end of Chapter 7 I have inserted a version of von Staudt's proof that the Desargues configuration (unlike the general Pappus configuration) it not merely self-dual but self-polar. The new Exercise 5 on page 124 shows that there is a Desargues configuration whose ten points and ten lines have coordinates involving only 0, 1, and -1. This scheme is of special interest because, when these numbers are interpreted as residues modulo 5 (so that the geometry is $PG(2, 5)$, as in Chapter 10), the ten pairs of perspective triangles are interchanged by harmonic homologies, and therefore the whole configuration is invariant for a group of 5! projective collineations, appearing as permutations of the digits 1, 2, 3, 4, 5 used on page 27. (The general Desargues configuration has the same 5! automorphisms, but these are usually not expressible as collineations. In fact, the perspective collineation $OPQR \to OP'Q'R'$ considered on page 53 is not, in general, of period two.*) Finally, there is a new Section 12.9 on page

* This remark corrects a mistake in my *Twelve Geometric Essays* (Southern Illinois University Press, 1968), p. 129.

132, briefly indicating how the theory changes if the diagonal points of a quadrangle are collinear.

I wish to express my gratitude to many readers of the first edition who have suggested improvements; especially to John Rigby, who noticed some very subtle points.

H. S. M. COXETER

Toronto, Canada
May, 1973

Contents

Projective Geometry

CHAPTER ONE

Introduction

If Desargues, the daring pioneer of the seventeenth century, could have foreseen what his ingenious method of projection was to lead to, he might well have been astonished. He knew that he had done something good, but he probably had no conception of just how good it was to prove.

E. T. Bell (1883–1960)
(Reference **3**, p. 244)

1.1 What is Projective Geometry?

The plane geometry of the first six books of Euclid's *Elements* may be described as the geometry of lines and circles: its tools are the straight-edge (or unmarked ruler) and the compasses. A remarkable discovery was made independently by the Danish geometer Georg Mohr (1640–1697) and the Italian Lorenzo Mascheroni (1750–1800). They proved that nothing is lost by discarding the straight-edge and using the compasses alone.* For instance, given four points A, B, C, D, we can still construct the point where the lines AB and CD would meet if we had the means to draw them; but the actual procedure is quite complicated. It is natural to ask how much remains if we discard the compass instead, and use the straight-edge alone.† At a glance, it looks as if nothing at all will remain: we cannot even carry out the construction described in Euclid's first proposition. Is it possible to develop a geometry

* See Reference **6**, pp. 144–151, or Reference **8**, p. 79.
† See Reference **16**, pp. 41–43.

having no circles, no distances, no angles, no intermediacy (or "betweenness"), and no parallelism? Surprisingly, the answer is Yes; what remains is projective geometry: a beautiful and intricate system of propositions, simpler than Euclid's but not too simple to be interesting. The passage from axioms and "obvious" theorems to unexpected theorems will be seen to resemble Euclid's work in spirit, though not in detail.

This geometry of the straight-edge seems at first to have very little connection with the familiar derivation of the name *geometry* as "earth measurement." Though it deals with points, lines, and planes, no attempt is ever made to measure the distance between two points or the angle between two lines. It does not even admit the possibility that two lines in a plane might fail to meet by being "parallel."

We naturally think of a *point* as "position without magnitude" or "an infinitesimal dot," represented in a diagram by a material dot only just big enough to be seen. By a *line* we shall always mean a straight line of unlimited extent. Part of a line is reasonably well represented by a thin, tightly stretched thread, or a ray of light. A *plane* is a flat surface of unlimited extent, that is, a surface that contains, for any two of its points, the whole of the line joining them. Any number of points that lie on a line are said to be *collinear*. Any number of lines that pass through a point are said to be *concurrent*. Any number of points or lines (or both) that lie in a plane are said to be *coplanar*.

People who have studied only Euclidean geometry regard it as an obvious fact that two coplanar lines with a common perpendicular are *parallel*, in the sense that, however far we extend them, they will remain the same distance apart. By stretching our imagination we can conceive the possibility that this is merely a first approximation: that if we could extend them for millions or billions of miles we might find the lines getting closer together or farther apart. When we look along a straight railroad we get the impression that the two parallel rails meet on the horizon. Anyhow, by assuming that two coplanar lines always meet, we obtain a system of propositions which (as we shall verify in Chapter 11) is just as logically consistent as Euclid's different system. In the words of D. N. Lehmer (Reference **12,** p. 12):

> As we know nothing experimentally about such things, we are at liberty to make any assumptions we please, so long as they are consistent and serve some useful purpose.

1.2 Historical Remarks

The motivation for this kind of geometry came from the fine arts. It was in 1425 that the Italian architect Brunelleschi began to discuss the geometrical theory of perspective, which was consolidated into a treatise by Alberti a few years later. Because of this application, it is natural to begin the subject in three-dimensional space; but we soon find that what happens in a single plane is sufficiently exciting to occupy our attention for a long time. Plane projective geometry may be described as the study of geometrical properties that are unchanged by "central projection," which is essentially what happens when an artist draws a picture of a tiled floor on a vertical canvas. The square tiles cease to be square, as their sides and angles are distorted by foreshortening; but the lines remain straight, since they are sections (by the picture-plane) of the planes that join them to the artist's eye. Thus projective geometry deals with triangles, quadrangles, and so on, but not with right-angled triangles, parallelograms, and so on. Again, when a lamp casts a shadow on a wall or on the floor, the circular rim of a lampshade usually casts a large circular or elliptic shadow on the floor and a hyperbolic shadow on the nearest wall. (Such "conic sections" or *conics* are sections of the cone that joins the source of light to the rim of the lampshade.) Thus projective geometry waives the customary distinction between a circle, an ellipse, a parabola, and a hyperbola; these curves are simply conics, all alike.

Although conics were studied by Menaechmus, Euclid, Archimedes and Apollonius, in the fourth and third centuries B.C., the earliest truly projective theorems were discovered by Pappus of Alexandria in the third century A.D., and it was J. V. Poncelet (1788–1867) who first proved such theorems by purely projective reasoning.

More than two hundred years before Poncelet, the important concept of a point at infinity occurred independently to the German astronomer Johann Kepler (1571–1630) and the French architect Girard Desargues (1591–1661). Kepler (in his *Paralipomena in Vitellionem*, 1604) declared that a parabola has two foci, one of which is infinitely distant in both of two opposite directions, and that any point on the curve is joined to this "blind focus" by a line parallel to the axis. Desargues (in his *Brouillon project . . .*, 1639) declared that parallel lines "*sont entre elles d'une mesme ordonnance dont le but est à distance infinie*." (That is, parallel lines have a common end at an infinite distance.) And again, "*Quand en un plan, aucun des points d'une droit n'y est à distance finie, cette droit y est à distance infinie*." (When no point of a line is at a finite distance, the line itself is at an infinite distance). The groundwork was thus laid for Poncelet to derive projective space from ordinary space by postulating a common "line at infinity" for all the planes parallel to a given plane. This ingenious device, which we shall analyze carefully in Chapter 11, serves to justify our assumption that, in a plane, any two lines meet; for, if the lines have no ordinary point in common, we say that they meet in a point at infinity. But we are not really working in projective geometry until we are prepared to forget the inferior status of such extra points and admit them into the community as full members having the same privileges as ordinary points. This emancipation of the subject was carried out by another German, K. G. C. von Staudt (1798–1867). The last vestiges of dependence on ordinary geometry were removed in 1871, when Felix Klein provided an algebraic foundation for projective geometry in terms of "homogeneous coordinates," which had been discovered independently by K. W. Feuerbach and A. F. Möbius in 1827.

The determination of a point by two lines nicely balances the determination of a line by two points. More generally, we shall find that every statement about points and lines (in a plane) can be replaced by a *dual* statement about lines and points. The possibility of making such a replacement is known as the "principle of duality." Poncelet claimed this principle as his own discovery; but its nature was more clearly understood by another Frenchman, J. D. Gergonne (1771–1859). Duality gives projective geometry a peculiar charm, making it more symmetrical than ordinary (Euclidean) geometry.

Besides being a thing of beauty in its own right, projective geometry is useful as supplying a fresh approach to Euclidean geometry. This is especially evident in the theory of conics, where a single projective theorem may yield several Euclidean theorems by different choices of the line at infinity; e.g., if the line at infinity is a tangent or a secant, the conic is a parabola or a hyperbola, respectively. Arthur Cayley (1821–1895) and Felix Klein (1849–1925) noticed that projective geometry is equally powerful in its application to non-Euclidean geometries. With characteristic enthusiam, the former said:

> Metrical geometry is a part of descriptive geometry, and descriptive geometry is all geometry.

(Cayley, in 1859, used the word "descriptive" where today we would say "projective.")

EXERCISES

1. Which of the following figures belong to projective geometry:
   (i) a parallelogram,
   (ii) an isosceles triangle,
   (iii) a triangle and its medians,
   (iv) a figure consisting of 4 points, no 3 collinear, and the lines joining them in pairs,
   (v) a circle with a diameter,
   (vi) a conic with a secant (i.e., a line meeting it twice),
   (vii) a plane curve with a tangent,
   (viii) a hexagon (consisting of 6 points named in cyclic order, and the 6 lines that join consecutive pairs)?

2. How could a lampshade be tilted so that its circular rim would yield a *parabolic* shadow on the wall?
3. Translate the following statements into the language of points at infinity:
   (i) Through a given point there passes just one line parallel to a given line.
   (ii) If two lines are parallel to a third line, they are parallel to each other.

1.3 Definitions

It is convenient to regard a line as a certain set of points, and a plane as a certain set of points and lines. A point and a line, or a point and a plane, or a line and a plane, are said to be *incident* if the former belongs to the latter. We also say that the former *lies on* (or in) the latter, and that the latter *passes through* the former. We shall consistently use capital italic letters for points, small (lower case) italic letters for lines, and Greek letters for planes. If a line l passes through two points P and Q, we say that it *joins* them and write $l = PQ$. Similarly, if a plane α passes through two lines l and m, or through l and a nonincident point P, we say that α *joins* the two lines, or the line and point, and write

$$\alpha = lm = ml = lP = Pl.$$

If P lies on both l and m, we say that these lines *meet* in P, or that P is their *common point* (or "intersection"):

$$P = l \cdot m.$$

(Notice the special use of the dot: lm is a plane, but $l \cdot m$ is a point.) Similarly, a line and a plane may have a common point $l \cdot \alpha$, and two planes may have a *common line* $\alpha \cdot \beta$.

J. L. Synge (Reference **18,** p. 32) has described an amusing and instructive game called Vish (short for "vicious circle"):

The Concise Oxford Dictionary devotes over a column to the word "point" . . . "that which has position but not magnitude." This definition passes the buck, as all definitions do. You now have to find out what "position" and "magnitude" are. This means further consultation of the Dictionary, and we may as well make the best of it by turning it into a game of Vish. So here goes.

Point = that which has position but not magnitude.
Position = place occupied by a thing.
Place = part of space. . . .
Space = continuous extension. . . .
Extension = extent.
Extent = space over which a thing extends.
Space = continuous extension. . . .

The word Space is repeated. We have Vish In Seven. . . . Well, what about it? Didn't we see and prove that a vicious circle is inevitable, so why be surprised that we get one here? If that is your reaction, I shout with joy. . . .

Vish illustrates the important principle that any definition of a word must inevitably involve other words, which require further definitions. The only way to avoid a vicious circle is to regard certain *primitive concepts* as being so simple and obvious that we agree to leave them undefined. Similarly, the proof of any statement uses other statements; and since we must begin somewhere, we agree to leave a few simple statements unproved. These primitive statements are called *axioms*.

In addition to the primitive concepts and axioms, we take for granted the words of ordinary speech, the ideas of logical argument, and the principle of one-to-one correspondence. The last is well illustrated by the example of cups and saucers. Suppose we had about a hundred cups and about a hundred saucers and wished to know whether the number of cups was actually equal to the number of saucers. This could be done, without counting, by the simple device of putting each cup on a saucer, that is, by establishing a one-to-one correspondence between the cups and saucers.

EXERCISES

1. Play Vish beginning with the words:
   (i) Axiom,
   (ii) Dimension,
   (iii) Fraction.

2. Set up a one-to-one correspondence between the sequence of natural numbers 1, 2, 3, 4, . . . and the sequence of even numbers 2, 4, 6, 8, ... Are we justified in saying that there are just as many even integers as there are integers altogether?

1.4 The Simplest Geometric Objects

A basis for projective geometry may be chosen in various ways. It seems simplest to use three primitive concepts: *point*, *line*, and *incidence*. In terms of these we can easily define "lie on," "pass through," "join," "meet," "collinear," "concurrent," and so on. It is not quite so obvious that we can define a plane; but if a point P and a line l are not incident, the *plane* Pl may be taken to consist of all the points that lie on lines joining P to points on l, and all the lines that join pairs of distinct points so constructed.

A *triangle* PQR consists of three noncollinear points P, Q, R, called its *vertices*, and the three joining lines QR, RP, PQ, called its *sides*. (When we have formulated the axioms and some of their simple consequences, we shall see that the triangle PQR can be proved to lie in the plane PQR.) Thus, if 3 points are joined in pairs by 3 lines, they form a triangle, which is equally well formed by 3 lines meeting by pairs in 3 points. The case of 4 points or 4 lines is naturally more complicated, and we will find it convenient to give the definitions in "parallel columns" (although it is not seriously expected that anybody will read the left column with the left eye and simultaneously the right column with the right eye).

If 4 points in a plane are joined in pairs by 6 distinct lines, they are called the vertices of a *complete quadrangle*, and the lines are its 6 sides. Two sides are said to be *opposite* if their common point is not a vertex. The common point of two opposite sides is called a *diagonal point*. There are 3 diagonal points. In Figure 1.4A, the quadrangle is $PQRS$, its sides are

$$PS, \quad QS, \quad RS,$$
$$QR, \quad RP, \quad PQ,$$

and its diagonal points are

$$A, \quad B, \quad C.$$

If 4 lines in a plane meet by pairs in 6 distinct points, they are called the sides of a *complete quadrilateral*, and the points are its 6 vertices. Two vertices are said to be *opposite* if their join is not a side. The join of two opposite vertices is called a *diagonal line*. There are 3 diagonal lines. In Figure 1.4A, the quadrilateral is $pqrs$, its vertices are

$$p \cdot s, \quad q \cdot s, \quad r \cdot s,$$
$$q \cdot r, \quad r \cdot p, \quad p \cdot q,$$

and its diagonal lines are

$$a, \quad b, \quad c.$$

When there is no possibility of misunderstanding, we speak simply of quadrangles and quadrilaterals, omitting the word "complete." This word was

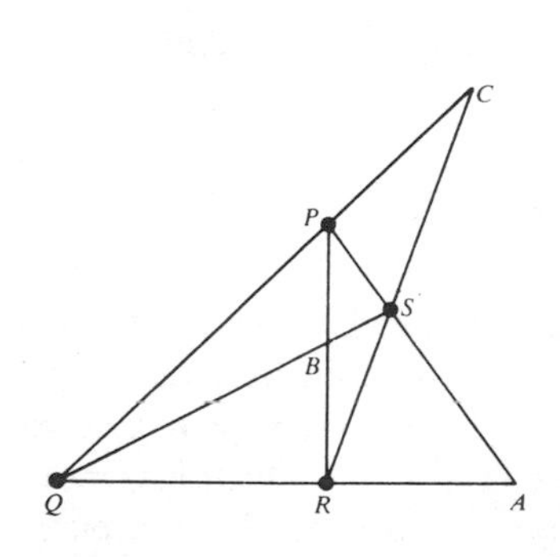

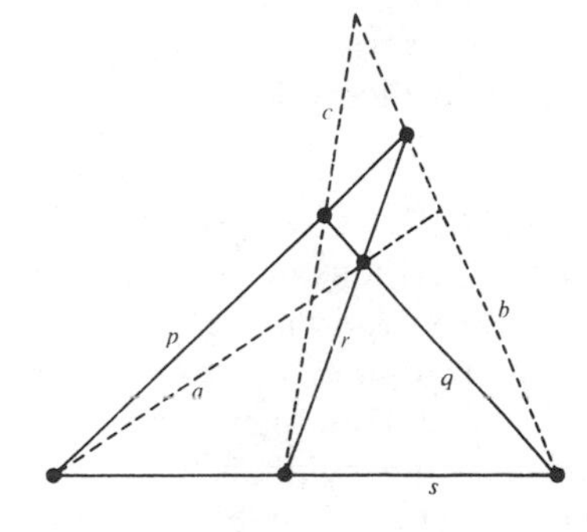

FIGURE 1.4A

introduced to avoid confusion with an ordinary quadrangle, which has 4 vertices and 4 sides; for instance, the ordinary quadrangle $PQRS$ has sides PQ, QR, RS, SP. It is more usual to call this a "quadrilateral," but to do so is unreasonable, as the word "triangle" refers to its vertices rather than its sides, and so too does the word "pentagon." The only other polygon that we shall have occasion to use is the (ordinary) *hexagon*, which has 6 vertices and 6 sides.

EXERCISES

Regarding the triangle as a complete 3-point and the complete quadrangle as a complete 4-point, define analogously (for any natural number n):

(i) a complete n-point, (ii) a complete n-line.

1.5 Projectivities

It is sometimes convenient to use the name *range* for the set of all points on a line, and *pencil* for the set of all lines that lie in a plane and pass through a point. Ranges and pencils are instances of *one-dimensional forms*. We shall often have occasion to consider a one-to-one correspondence between two one-dimensional forms. The simplest such correspondence between a range and a pencil arises when corresponding members are incident. In this case it is naturally understood that the line o on which the points of the range lie is *not* incident with the point O through which the lines of the pencil pass. Thus the range is a *section* of the pencil (namely, the section by the line o) and the pencil *projects* the range (from the point O). As a notation for this *elementary* correspondence we may write either

$$X \barwedge x,$$

where X is a variable point of the range and x is the corresponding line of the pencil (as in Figure 1.5A), or

$$ABC \cdots \barwedge abc \cdots,$$

where $A, B, C, \ldots$ are particular positions of X and $a, b, c, \ldots$ are the corresponding positions of x (as in Figure 1.5B).

In such a relation, the order in which the symbols for the points or lines are written does not necessarily agree with the order in which the points or lines occur in the range or pencil. (In fact, the latter "order" is not defined!) Corresponding symbols are placed in corresponding positions, but the statement $ABC \cdots \barwedge abc \cdots$, has the same meaning as $BAC \cdots \barwedge bac \cdots$, and so forth.

Since the statement $X \barwedge x$ means that X and x are incident, we can just as well write

$$x \barwedge X;$$

but now it is convenient to make a subtle distinction. The correspondence $X \barwedge x$ is directed "from X to x": it *transforms* X into x; but the *inverse* correspondence $x \barwedge X$ transforms x into X.

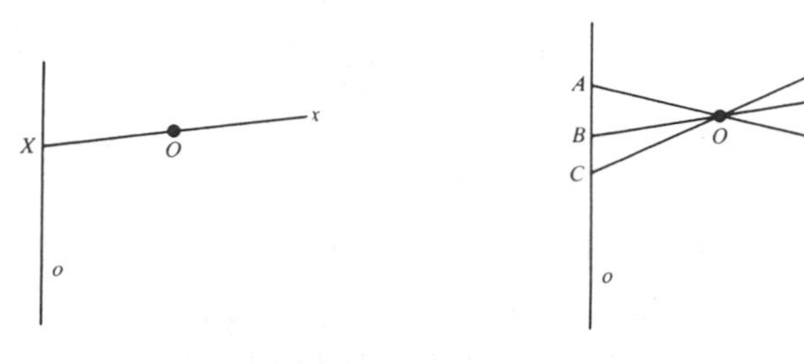

FIGURE 1.5A FIGURE 1.5B

A more sophisticated kind of transformation can be constructed by combining any number of elementary correspondences. For this purpose, we use a sequence of lines and points occurring alternately:

$$o, O, o_1, O_1, o_2, \ldots, O_{n-1}, o_n, O_n.$$

We allow the sequence to begin with a point (by omitting o) or to end with a line (by omitting O_n, as in Figure 1.5C), but we insist that adjacent members

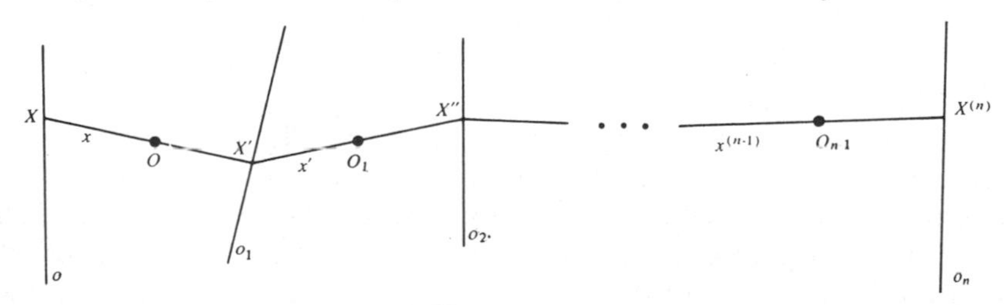

FIGURE 1.5C

shall be nonincident and that alternate members (such as O and O_1, or o_1 and o_2) shall be distinct. This arrangement of lines and points enables us to establish a transformation relating the range of points X on o (or the pencil of lines x through O) to the pencil of lines $x^{(n)}$ through O_n (or the range of points $X^{(n)}$ on o_n). We call such a transformation a *projectivity*.

Instead of

$$X \barwedge x \barwedge X' \barwedge x' \barwedge X'' \barwedge \cdots \barwedge X^{(n)} \barwedge x^{(n)}$$

we write simply

$$X \barwedge x^{(n)}$$

or $\quad x \barwedge x^{(n)}, \quad$ or $\quad x \barwedge X^{(n)}, \quad$ or $\quad X \barwedge X^{(n)}.$

In other words, we extend the meaning of the sign $\barwedge$ from an elementary correspondence to the *product* (or "resultant") of any number of elementary correspondences.

This extension of meaning is comparable to the stage in arithmetic when we extend the meaning of *number* from an integer to a fraction: the quotient of two integers.

EXERCISES

1. In the elementary correspondence $X \barwedge x$, why is it necessary for the line o and the point O to be nonincident?
2. Draw a version of Figure 1.5c using three points A, B, C, as in Figure 1.5B.
3. Draw a version of Figure 1.5c with $n = 2$ and $o_2 = o$, so as to establish a projectivity $X \barwedge X''$ relating pairs of points on o. Where is X'' when X is on o_1? Where is X'' when X is on OO_1?

1.6 Perspectivities

One kind of projectivity is sufficiently important to deserve a special name and a slightly more elaborate sign: the product of *two* elementary correspondences is called a *perspectivity* and is indicated by the sign $\doublebarwedge$ (with *two* bars). Using Poncelet's device of parallel columns to emphasize the "principle of duality," as in Section 1.4, we may describe this transformation as follows:

Two ranges are related by a perspectivity with *center* O if they are sections of one pencil (consisting of all the lines through O) by two distinct lines o and o_1; that is, if the join XX' of corresponding points continually passes through the point O. In symbols:

$$X \doublebarwedge X' \quad \text{or} \quad X \overset{O}{\doublebarwedge} X'.$$

Two pencils are related by a perspectivity with *axis* o_1 if they project one range (consisting of all the points on o_1) from two distinct points O and O_1; that is, if the intersection $x \cdot x'$ of corresponding lines continually lies on the line o_1. In symbols:

$$x \doublebarwedge x' \quad \text{or} \quad x \overset{o_1}{\doublebarwedge} x'.$$

For instance, in Figure 1.6A (where A, B, C are particular instances of the variable point X, and a, b, c of the variable line x), we have the perspectivities

$$ABC \overset{O}{\doublebarwedge} A'B'C', \qquad abc \overset{o_1}{\doublebarwedge} a'b'c',$$

which can be analyzed in terms of elementary correspondences as follows:

$$ABC \barwedge abc \barwedge A'B'C', \qquad abc \barwedge A'B'C' \barwedge a'b'c'.$$

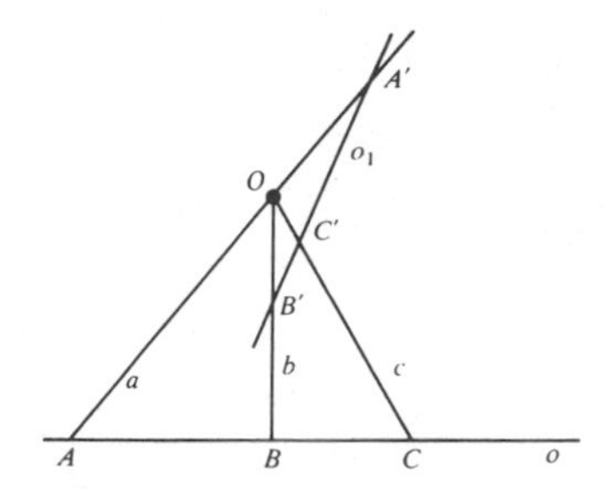

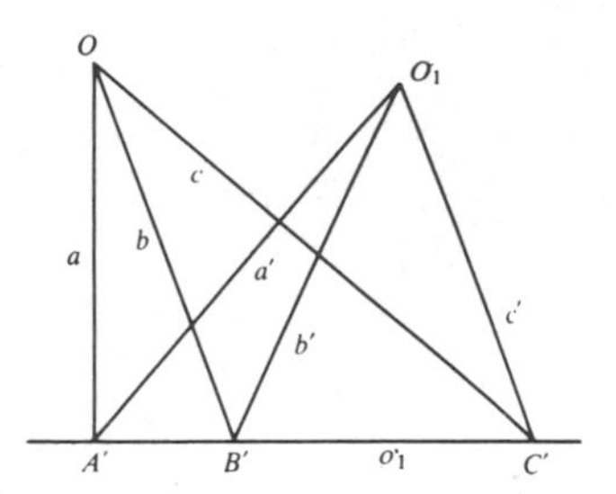

FIGURE 1.6A

Given three distinct points A, B, C on a line, and three distinct points A'', B'', C'' on another line, we can set up two perspectivities whose product has the effect

$$ABC \barwedge A''B''C''$$

in the manner of Figure 1.6B, where the *axis* (or "intermediary line") of the projectivity joins the points

$$B' = AB'' \cdot BA'', \qquad C' = AC'' \cdot CA'',$$

so that if $A' = AA'' \cdot B'C'$,

(1.61) $$ABC \overset{A''}{\barwedge} A'B'C' \overset{A}{\barwedge} A''B''C''.$$

For each point X on AB, we can construct a corresponding point X'' on $A''B''$ by joining A to the point $X' = A''X \cdot B'C'$, so that

(1.62) $$ABCX \overset{A''}{\barwedge} A'B'C'X' \overset{A}{\barwedge} A''B''C''X''.$$

We shall see, in Chapter 4, that this projectivity $ABC \barwedge A''B''C''$ is unique, in the sense that *any* sequence of perspectivities relating ABC to $A''B''C''$ will have the same effect on X.

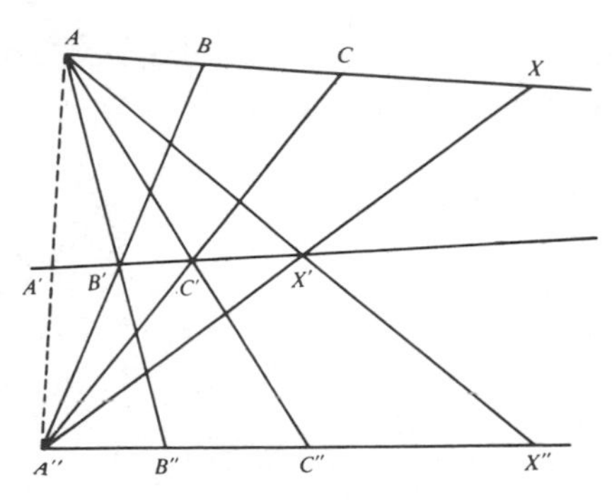

FIGURE 1.6B

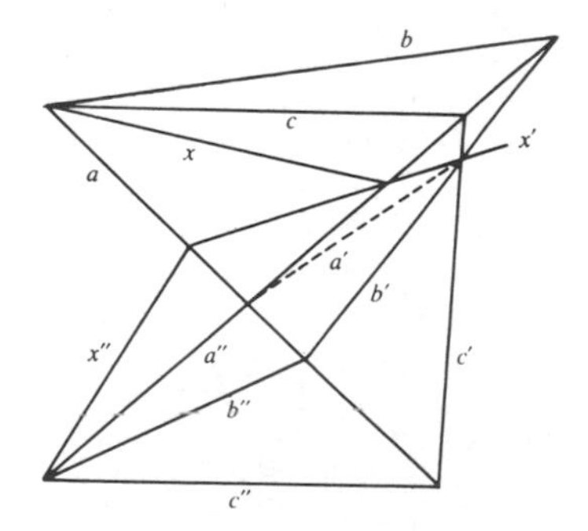

FIGURE 1.6C

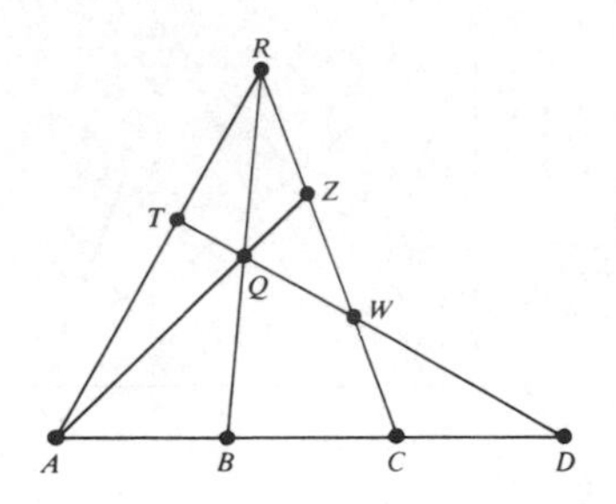

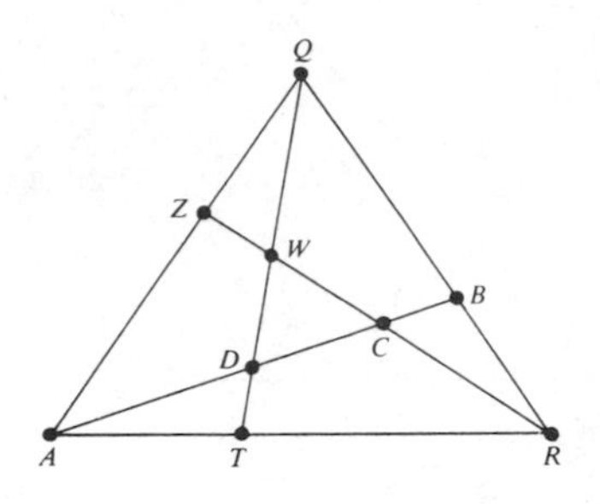

FIGURE 1.6D

Interchanging points and lines, we obtain an analogous construction (Figure 1.6C) for the projectivity $abc \barwedge a''b''c''$, where a, b, c are three distinct lines through a point and a'', b'', c'' are three distinct lines through another point.

Another example of a projectivity is illustrated in Figure 1.6D, where A, B, C, D are any four collinear points, R is a point outside their line, T, Q, W are the sections of RA, RB, RC by an arbitrary line through D, and Z is the point $AQ \cdot RC$. In this case

$$ABCD \overset{Q}{\barwedge} ZRCW \overset{A}{\barwedge} QTDW \overset{R}{\barwedge} BADC.$$

Hence

$$ABCD \barwedge BADC.$$

Expressing this result in words, we have the following theorem:

1.63 *Any four collinear points can be interchanged in pairs by a projectivity.*

As a third instance, we have, in Figure 1.6E,

$$ABC \overset{R}{\barwedge} APS \overset{Q}{\barwedge} AFB, \qquad ABC \overset{S}{\barwedge} AQR \overset{P}{\barwedge} AFB.$$

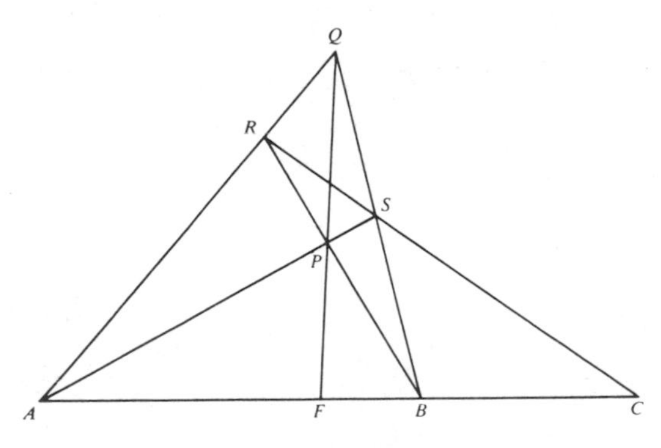

FIGURE 1.6E

In this projectivity $ABC \barwedge AFB$, the point A corresponds to itself. A point that corresponds to itself is said to be *invariant*.

The idea of a projectivity is due to Poncelet. Its analysis into elementary correspondences was suggested by Mathews (Reference **14,** p. 39). The sign $\barwedge$ was invented by von Staudt. For the special case of a perspectivity, the sign $\doublebarwedge$ was adopted by the great American geometer Oswald Veblen (1880–1960).

EXERCISES

1. Given three collinear points A, B, C, set up two perspectivities whose product has the effect $ABC \barwedge BAC$.

2. Given three concurrent lines a, b, c set up two perspectivities whose product has the effect $abc \barwedge bac$.

3. Given three collinear points A, B, C and three concurrent lines a, b, c, set up five elementary correspondences ("two-and-a-half perspectivities") whose product has the effect $ABC \barwedge abc$.

4. Given four collinear points A, B, C, D, set up three perspectivities whose product has the effect $ABCD \barwedge DCBA$.

CHAPTER TWO

Triangles and Quadrangles

> To construct a geometry is to state a system of axioms and to deduce all possible consequences from them. All systems of pure geometry . . . are constructed in just this way. Their differences . . . are differences not of principle or of method, but merely of richness of content and variety of application . . . . You must naturally be prepared to sacrifice simplicity to some extent if you wish to be interesting.
>
> *G. H. Hardy* (1877–1947)
> ("What is geometry?" *Mathematical Gazette*, **12** (1925), pp. 314, 315)

2.1 Axioms

As we saw in Section 1.3, the complete development of any branch of mathematics must begin with some undefined entities (primitive concepts) and unproved propositions (axioms). The precise choice is a matter of taste. It is, of course, essential that the axioms be consistent (not contradicting one another) and it is desirable that they be independent, simple, and plausible. Such foundations for projective geometry were first proposed by two Italians: Gino Fano (in 1892) and Mario Pieri (in 1899). The following eight axioms, involving three primitive concepts (*point*, *line*, and *incidence*) differ only slightly from those proposed by Veblen and Young (Reference **19**. pp. 16, 18, 24, 45). (We have already seen how the words *plane*, *quadrangle*, and *projectivity* can be defined in terms of the primitive concepts.)

AXIOM **2.11** *There exist a point and a line that are not incident.*

AXIOM **2.12** *Every line is incident with at least three distinct points.*

AXIOM **2.13** *Any two distinct points are incident with just one line.*

AXIOM **2.14** *If A, B, C, D are four distinct points such that AB meets CD, then AC meets BD.* (See Figure 2.1A.)

AXIOM **2.15** *If ABC is a plane, there is at least one point not in the plane ABC.*

AXIOM **2.16** *Any two distinct planes have at least two common points.*

AXIOM **2.17** *The three diagonal points of a complete quadrangle are never collinear.* (See Figure 1.4A.)

AXIOM **2.18** *If a projectivity leaves invariant each of three distinct points on a line, it leaves invariant every point on the line.*

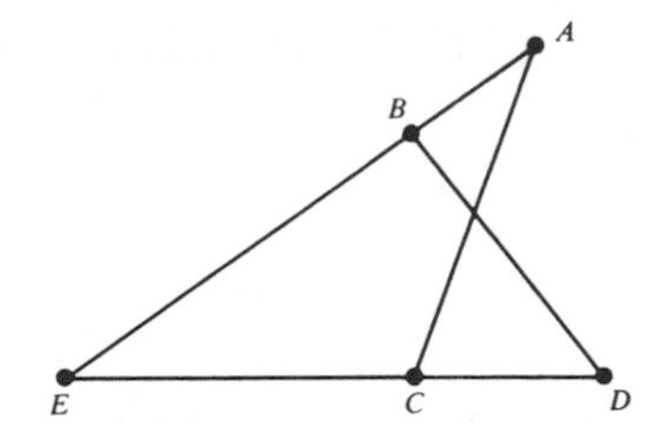

FIGURE 2.1A

The best possible advice to the reader is to set aside all his previously acquired knowledge (such as trigonometry and analytic geometry) and use only the axioms and their consequences. He may occasionally be tempted to use the old methods to work out one of the exercises; but then he is likely to be so engulfed in ugly calculations that he will return to the synthetic method with renewed enthusiasm.

EXERCISES

1. Give detailed proofs of the following theorems, pointing out which axioms are used:
   (i) There exist at least four distinct points.
   (ii) If a is a line, there exists a point not lying on a.
   (iii) If A is a point, there exists a line not passing through A.
   (iv) Every point lies on at least three lines.
2. Construct a projectivity having exactly two invariant points. [*Hint:* Use Exercise 3 of Section 1.5.]

3. Draw an equilateral triangle ABC with its incircle DEF, medians AD, BE, CF, and center G. Notice that the figure involves 7 points, 6 lines, and 1 circle. Consider a "geometry," consisting entirely of 7 points and 7 lines, derived from the figure by calling the circle a line (and ignoring the extra intersections). Which one of the "two-dimensional" axioms (Axioms 2.11, 2.12, 2.13, 2.14, 2.17, 2.18) is denied? [*Hint:* Where are the diagonal points of the quadrangles $ABCG$, $AEFG$, $BCEF$?]

2.2 Simple Consequences of the Axioms

Most readers will have no difficulty in accepting Axioms 2.11, 2.12, 2.13. The first departure from Euclidean geometry appears in Axiom 2.14, which rules out the possibility that AC and BD might fail to meet by being "parallel." This axiom, which resembles Pasch's Axiom (12.27 of Reference **8,** p. 178), is Veblen's ingenious device for declaring that any two *coplanar* lines have a common point *before* defining a plane! (In fact, the line BD, whose intersection with AC is asserted, must lie in the plane AEC, where $E = AB \cdot CD$, since B lies on AE, and D on EC.)

Given a triangle ABC, we can define a *pencil* of lines through C as consisting of all the lines CX, where X belongs to the range of points on AB. The first four axioms are all that we need in order to define the *plane* ABC as a certain set of points and lines, namely, all the points on all the lines of the pencil, and all the lines that join pairs of such points. We then find that the same plane is determined when we replace C by another one of the points, and AB by one of the lines not incident with this point.

Axiom 2.15 makes the geometry three-dimensional, and Axiom 2.16 prevents it from being four-dimensional. (In fact, four-dimensional geometry would admit a pair of planes having only one common point!) It follows that two distinct planes, α and β, meet in a line, which we call the line $\alpha \cdot \beta$.

In virtue of Axiom 2.17, the diagonal points of a quadrangle form a triangle. This is called the *diagonal triangle* of the quadrangle. It will be found to play an important role in some of the later developments. However (as we shall see in the exercise at the end of Section 10.3), there are some interesting, though peculiar, geometries in which the diagonal points of a quadrangle are always collinear, so that Axiom 2.17 is denied. (See, e.g., Exercise 3 below.)

The plausibility of Axiom 2.18 will appear in Section 3.5, where we shall prove, on the basis of the remaining seven axioms, that a projectivity having three invariant points leaves invariant, if not the whole line, so many of its points that they have the "appearance" of filling the whole line.

As an indication of the way axioms lead to theorems, let us now state four simple theorems and give their proofs in detail.

2.21 *Any two distinct lines have at most one common point.*

PROOF. Suppose, if possible, that two given lines have two common points A and B. Axiom 2.13 tells us that each line is determined by these two points. Thus the two lines coincide, contradicting our assumption that they are distinct.

2.22 *Any two coplanar lines have at least one common point.*

PROOF. Let E be a point coplanar with the two lines but not on either of them. Let AC be one of the lines. Since the plane ACE is determined by the pencil of lines through E that meet AC, the other one of the two given lines may be taken to join two points on distinct lines of this pencil, say B on EA, and D on EC, as in Figure 2.1A. According to Axiom 2.14, the two lines AC and BD have a common point.

2.23 *If two lines have a common point, they are coplanar.*

PROOF. If two lines have a common point C, we may name them AC, BC, and conclude that they lie in the plane ABC.

2.24 *There exist four coplanar points of which no three are collinear.*

PROOF. By our first three axioms, there exist two distinct lines having a common point and each containing at least two other points, say lines EA and EC containing also B and D, respectively, as in Figure 2.1A. The four distinct points A, B, C, D have the desired property of noncollinearity. For instance, if the three points A, B, C were collinear, E (on AB) would be collinear with all of them, and EA would be the same line as EC, contradicting our assumption that these two lines are distinct.

Without Theorem 2.24, Axiom 2.17 might be "vacuous": it merely says that, if a complete quadrangle exists, its three diagonal points are not collinear.

Notice the remarkably compact foundation which is now seen to suffice for the erection of the whole system of projective geometry.

EXERCISES

1. Prove the following theorems:
   (i) There exist four coplanar lines of which no three are concurrent.
   (ii) The three diagonal lines of a complete quadrilateral are never concurrent. (They are naturally said to form the *diagonal triangle* of the quadrilateral.)
2. Draw complete quadrangles and quadrilaterals of various shapes, indicating for each its diagonal triangle.

2.3 Perspective Triangles

Two ranges or pencils are said to be *perspective* if they are related by a perspectivity. This notion can be extended to plane figures involving more than one point and more than one line, as follows. Two specimens of such a figure are said to be *perspective* if their points can be put into one-to-one correspondence so that pairs of corresponding points are joined by concurrent lines, or if their lines can be put into one-to-one correspondence so that pairs of

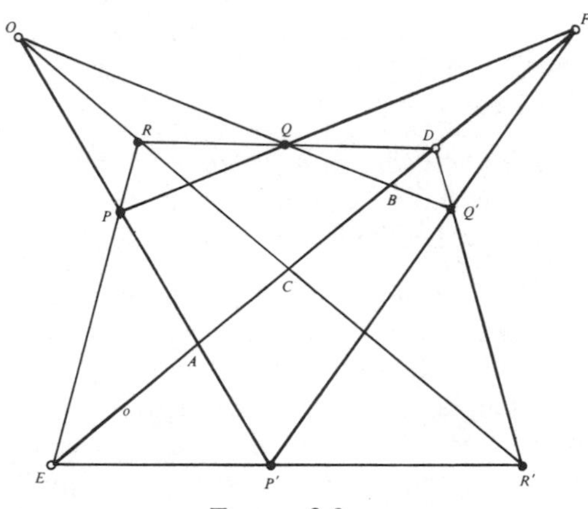

FIGURE 2.3A

corresponding lines meet in collinear points. For instance, the two triangles PQR and $P'Q'R'$ in Figure 2.3A are perspective since corresponding vertices are joined by the three concurrent lines PP', QQ', RR', or since corresponding sides meet in the three collinear points

$$D = QR \cdot Q'R', \qquad E = RP \cdot R'P', \qquad F = PQ \cdot P'Q'.$$

When Theorems 2.31 and 2.32 have been stated and proved, we shall see that either kind of correspondence implies the other. Meanwhile, let us say tentatively that two figures are *perspective from a point O* if pairs of corresponding points are joined by lines through O, and that two figures are *perspective from a line o* if pairs of corresponding lines meet on o. (It is sometimes convenient to call O the *center*, and o the *axis*. Whenever we speak of perspective figures we assume that the points, and also the lines, are all distinct; for example, in the case of a pair of triangles, we assume that there are six distinct vertices and six distinct sides.) The desired identification will follow for all more complicated figures as soon as we have established it for triangles. Accordingly, we begin with the following theorem:

2.31 *If two triangles are perspective from a line they are perspective from a point.*

PROOF. Let two triangles, PQR and $P'Q'R'$, be perspective from a line o. In other words, let o contain three points D, E, F, such that D lies on both QR and $Q'R$, E on both RP and $R'P'$, F on both PQ and $P'Q'$. We wish to prove that the three lines PP', QQ', RR' all pass through one point O, as in Figure 2.3A. We distinguish two cases, according as the given triangles are in distinct planes or both in one plane.

(1) According to Axiom 2.14, since QR meets $Q'R'$, QQ' meets RR'. Similarly RR' meets PP', and PP' meets QQ'. Thus the three lines PP', QQ', RR' all meet one another. If the planes PQR and $P'Q'R'$ are distinct, the three lines must be concurrent; for otherwise they would form a triangle, and this triangle would lie in both planes.

(2) If PQR and $P'Q'R'$ are in one plane, draw, in another plane through o, three nonconcurrent lines through D, E, F, respectively, so as to form a triangle $P_1Q_1R_1$, with Q_1R_1 through D, R_1P_1 through E, and P_1Q_1 through F. This triangle is perspective from o with both PQR and $P'Q'R'$. By the result for noncoplanar triangles, the three lines PP_1, QQ_1, RR_1 all pass through one point S, and the three lines $P'P_1$, $Q'Q_1$, $R'R_1$ all pass through another point S'. (The points S and S' are distinct; for otherwise P_1 would lie on PP' instead of being outside the original plane.) Since P_1 lies on both PS and $P'S'$, Axiom 2.14 tells us that SS' meets PP'. Similarly SS' meets both QQ' and RR'. Hence, finally, the three lines PP', QQ', RR' all pass through the point

$$O = PQR \cdot SS'.$$

The converse is

2.32 DESARGUES'S THEOREM. *If two triangles are perspective from a point they are perspective from a line.*

PROOF. Let two triangles, PQR and $P'Q'R'$ (coplanar or noncoplanar) be perspective from a point O. We see from Axiom 2.14 that their three pairs of corresponding sides meet, say in D, E, F. It remains to be proved that these three points are collinear, as in Figure 2.3A. Consider the two triangles $PP'E$ and $QQ'D$. Since pairs of corresponding sides meet in the three collinear points R', R, O, these triangles are perspective from a line, and therefore (by 2.31), perspective from a point, namely, from the point $PQ \cdot P'Q' = F$. That is, the three points E, D, F are collinear.

Theorem 2.31, the converse of Desargues's theorem, happens to be easier to prove *ab initio* than Desargues's theorem itself. If, instead, we had proved 2.32 first (as in Reference **7**, p. 8), we could have deduced the converse by applying 2.32 to the triangles $PP'E$ and $QQ'D$.

EXERCISES

1. Verify experimentally the correctness of Desargues's theorem and its converse for perspective triangles of various shapes in various relative positions. If P, Q, R, P' and Q' are given, how much freedom do we have in choosing the position of R'?

2. If three triangles are all perspective from the same center, then* the three axes are concurrent. [*Hint:* Let the three axes be D_1E_1, D_2E_2, D_3E_3. Apply 2.31 to the triangles $D_1D_2D_3$ and $E_1E_2E_3$.]

3. What happens to Theorem 2.31 if we allow corresponding sides of the two triangles to be parallel, and admit points at infinity?

2.4 Quadrangular Sets

A *quadrangular set* is the section of a complete quadrangle by any line g that does not pass through a vertex. It is thus, in general, a set of six collinear points, one point on each side of the quadrangle; but the number of points is reduced to five or four if the line happens to pass through one or two diagonal points.

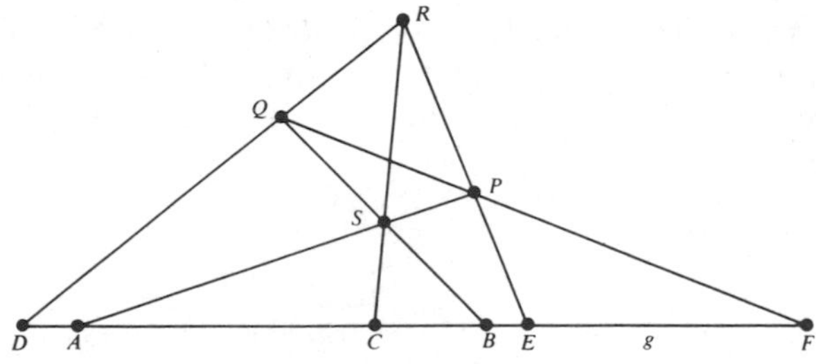

FIGURE 2.4A

Slightly changing the notation Q(ABC, DEF) of Veblen and Young (Reference **19**, p. 49), let us use the symbol

$$(AD)(BE)(CF)$$

to denote the statement that the six points A, B, C, D, E, F, form a quadrangular set in the manner of Figure 2.4A (that is, lying on the respective sides PS, QS, RS, QR, RP, PQ of the quadrangle), so that the first three points lie on sides through one vertex while the remaining three lie on the respectively opposite sides, which form a triangle. This statement is evidently unchanged if

* Whenever an exercise is phrased as a statement, we understand that it is a theorem to be proved. The omission of the words "prove that" or "show that" saves space.

we apply any permutation to ABC and the same permutation to DEF; for instance, $(AD)(BE)(CF)$ has the same meaning as $(BE)(AD)(CF)$, since the quadrangle $PQRS$ can equally well be called $QPRS$. Similarly, the statement $(AD)(BE)(CF)$ is equivalent to each of

$$(AD)(EB)(FC), \qquad (DA)(BE)(FC), \qquad (DA)(EB)(CF).$$

Any five collinear points A, B, C, D, E may be regarded as belonging to a quadrangular set. To see this, draw a triangle QRS whose sides RS, SQ, QR pass, respectively, through C, B, D. (These sides may be any three nonconcurrent lines through C, B, D.) We can now construct $P = AS \cdot ER$ and $F = g \cdot PQ$. If we had chosen a different triangle QRS, would we still have obtained the same point F? Yes:

2.41 *Each point of a quadrangular set is uniquely determined by the remaining points.*

PROOF. To show that F is uniquely determined by A, B, C, D, E (Figure 2.4B), we set up another quadrangle $P'Q'R'S'$ whose first five sides pass through the same five points on g. Since the two triangles PRS and $P'R'S'$ are perspective from g, Theorem 2.31 tells us that they are also perspective from a point; thus the line PP' passes through the point $O = RR' \cdot SS'$. Similarly, the perspective triangles QRS and $Q'R'S'$ show that QQ' passes through this same point O. (In other words, $PQRS$ and $P'Q'R'S'$ are perspective quadrangles.) By Theorem 2.32, the triangles PQR and $P'Q'R'$, which are perspective from the point O, are also perspective from the line DE, which is g; that is, the sides PQ and $P'Q'$ both meet g in the same point F.

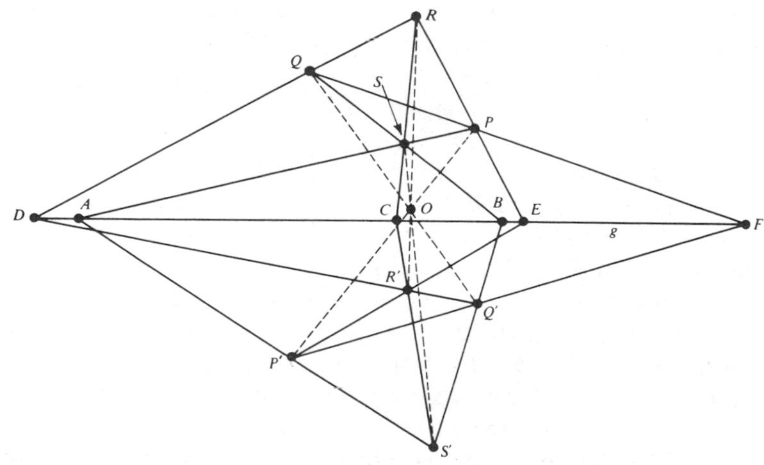

FIGURE 2.4B

EXERCISES

1. A necessary and sufficient condition for three lines containing the vertices of a triangle to be concurrent is that their sections A, B, C by a line g form, with the sections D, E, F of the sides of the triangle, a quadrangular set.
2. If, on a given transversal line, two quadrangles determine the same quadrangular set (in the manner of Figure 2.4B), their diagonal triangles are perspective.

2.5 Harmonic Sets

A *harmonic set* of four collinear points may be defined to be the special case of a quadrangular set when the line g joins two diagonal points of the

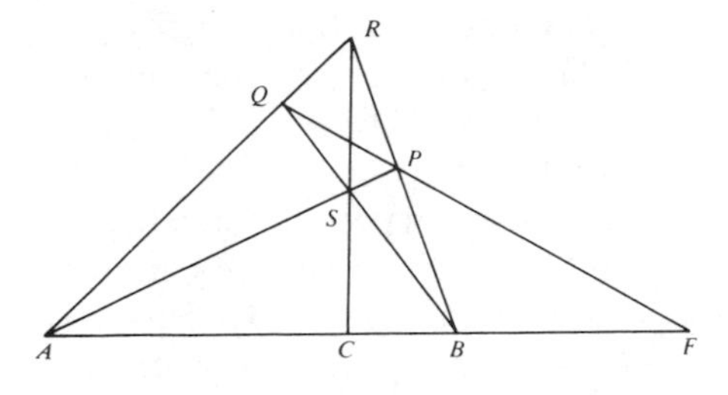

FIGURE 2.5A

quadrangle, as in Figure 2.5A or 1.6E. Because of the importance of this special case, we write the relation $(AA)(BB)(CF)$ in the abbreviated form

$$\mathsf{H}(AB, CF),$$

which evidently has the same meaning as $\mathsf{H}(BA, CF)$ or $\mathsf{H}(AB, FC)$ or $\mathsf{H}(BA, FC)$, namely that A and B are two of the three diagonal points of a quadrangle while C and F lie, respectively, on the sides that pass through the third diagonal point. We call F the *harmonic conjugate* of C with respect to A and B. Of course also C is the harmonic conjugate of F. From Theorem 2.41 we see that F is uniquely determined by A, B, C. (We shall prove in Theorem 3.35 that the relation $\mathsf{H}(AB, CF)$ implies $\mathsf{H}(CF, AB)$.) For a simple construction, draw a triangle QRS whose sides QR, QS, RS pass through A, B, C (as in Figure 2.5A); then $P = AS \cdot BR$ and $F = AB \cdot PQ$.

Axiom 2.17 implies that these harmonic conjugates C and F are distinct, except in the degenerate case when they coincide with A or B. In other words,

2.51 *If A, B, C are all distinct, the relation* $\mathsf{H}(AB, CF)$ *implies that F is distinct from C.*

It follows that there must be at least four points on every line; how many more is not specified. We shall examine this question in Section 3.5.

EXERCISES

1. Assigning the symbol G to $PQ \cdot RS$, name two other harmonic sets in Figure 2.5A.

2. How should the points P, Q, R, S in Figure 2.5A be renamed if the names of C and F were interchanged?

3. Derive a harmonic set from a quadrangle consisting of a triangle PQR and a point S inside.

4. What is the harmonic set determined by a quadrangle $PQRS$ if Axiom 2.17 is denied and the diagonal points are collinear?

5. Suppose, for a moment, that the projective plane is regarded as an extension of the Euclidean plane (as in Section 1.2). Referring to Figure 2.5A, suppose PQ is parallel to AB, so that F is at infinity. What metric result can you deduce about the location of C with reference to A and B?

6. Still working in the Euclidean plane, draw a line-segment OC, take G two-thirds of the way along it, and E two-fifths of the way from G to C. (For instance, make the distances in centimeters $OG = 10$, $GE = 2$, $EC = 3$.) If the segment OC represents a stretched string, tuned to the note C, the same string stopped at E or G will play the other notes of the major triad. By drawing a suitable quadrangle, verify experimentally that $\mathsf{H}(OE, CG)$. (Such phenomena explain our use of the word *harmonic*.)

CHAPTER THREE

The Principle of Duality

On November 18, 1812, the exhausted remnant of the French army . . . was overwhelmed at Krasnoi. Among those left for dead on the frozen battlefield was young Poncelet. . . . A searching party, discovering that he still breathed, took him before the Russian staff for questioning. As a prisoner of war . . . at Saratoff on the Volga . . . , he remembered that he had received a good mathematical education, and to soften the rigours of his exile he resolved to reproduce as much as he could of what he had learned. It was thus that he created projective geometry.

E. T. Bell (Reference **3**, pp. 238–239)

3.1 The Axiomatic Basis of the Principle of Duality

The geometry of points on a line is said to be one-dimensional. The geometry of points and lines in a plane is said to be two-dimensional. The geometry of points, lines, and planes in space is said to be three-dimensional. It is interesting to observe that the only place where we made essential use of three-dimensional ideas was in proving 2.31. This excursion "along the third dimension" could have been avoided by regarding Desargues's theorem, 2.32 (which implies 2.31) as a new axiom, replacing the three-dimensional axioms 2.15 and 2.16. For a purely two-dimensional theory, we can replace the three axioms 2.11, 2.12, 2.14 by the following two simpler statements:

3.11 *Any two lines are incident with at least one point.*

3.12 *There exist four points of which no three are collinear.*

(These are derived from 2.22 and 2.24 by omitting the word "coplanar," which is now superfluous.) We shall develop such a theory out of the following six propositions:

$$2.13, 3.11, 3.12, 2.17, 2.32, 2.18,$$

which will thus be seen to form a sufficient set of *axioms for the projective plane*.

The two-dimensional *principle of duality* (i.e., the principle of duality in the plane) asserts that every definition remains significant, and every theorem remains true, when we interchange the words *point* and *line* (and consequently also certain other pairs of words such as join and meet, collinear and concurrent, vertex and side, and so forth). For instance, the *dual* of the point $AB \cdot CD$ is the line $(a \cdot b)(c \cdot d)$. (Since duality interchanges joining and meeting, it requires not only the interchange of capital and small letters but also the removal of any dots that are present and the insertion of dots where they are absent.) The arrangement in Section 1.4 of definitions in "parallel columns" shows at once that the dual of a quadrangle (with its three diagonal points) is a quadrilateral (with its three diagonal lines). Still more simply, the dual of a triangle (consisting of its vertices and sides) is again a triangle (consisting of its sides and vertices); thus a triangle is an instance of a *self-dual* figure.

Axioms 2.13 and 3.11 clearly imply their duals. To prove the dual of 3.12, consider the sides PQ, QR, RS, SP of the quadrangle $PQRS$ that is given by 3.12 itself. Similarly, the duals of 2.17 and 2.18 present no difficulty. The dual of 2.32 is its converse, 2.31 (which can be proved by applying 2.32 to the triangles $PP'E$ and $QQ'D$ in Figure 2.3A). Thus *all the axioms for the projective plane imply their duals.*

This remark suffices to establish the validity of the two-dimensional principle of duality. In fact, after using the axioms and their consequences in proving a given theorem, we can immediately assert the dual theorem; for, a proof of the dual theorem could be written down quite mechanically by dualizing each step in the proof of the original theorem. (Of course, our proof of 2.31 cannot be dualized in this sense, because it is three-dimensional; but this objection is avoided by taking 2.32 as an axiom and observing that 2.31 can be deduced from it without either leaving the plane or appealing to the principle of duality.)

One of the most attractive features of projective geometry is the symmetry and economy with which it is endowed by the principle of duality: fifty detailed proofs may suffice to establish as many as a hundred theorems.

So far, we have considered only the *two-dimensional* principle of duality. But by returning to our original Axioms 2.11 through 2.18, we can just as easily establish the *three-dimensional* principle of duality, in which points, lines, and planes are interchanged with planes, lines, and points. For instance, if lines a and b are coplanar, the dual of ab is $a \cdot b$. However, for the sake of brevity we shall assume, until the end of Chapter 9, that all the points and lines considered are *in one plane*. (For a glimpse of the analogous three-dimensional developments, see Reference **8**, p. 256.)

EXERCISES

1. Deduce 2.12 from 2.13, 3.11, and 3.12 (Reference **8**, pp. 233, 446).
2. Let the diagonal points of a quadrangle $PQRS$ be

$$A = PS \cdot QR, \qquad B = QS \cdot RP, \qquad C = RS \cdot PQ,$$

as in Figure 1.4A. Define the further intersections

$$A_1 = BC \cdot QR, \qquad B_1 = CA \cdot RP, \qquad C_1 = AB \cdot PQ,$$

$$A_2 = BC \cdot PS, \qquad B_2 = CA \cdot QS, \qquad C_2 = AB \cdot RS.$$

Then the triads of points $A_1B_2C_2$, $A_2B_1C_2$, $A_2B_2C_1$, $A_1B_1C_1$ lie on lines, say p, q, r, s, forming a quadrilateral $pqrs$ whose three diagonal lines are the sides

$$a = BC, \qquad b = CA, \qquad c = AB$$

of the triangle ABC. In other words, the quadrangle $PQRS$ and the quadrilateral $pqrs$ have the same diagonal triangle. (Reference **19**, pp. 45–46.)

3.2 The Desargues Configuration

A set of m points and n lines in a plane is called a *configuration* (m_c, n_d) if c of the n lines pass through each of the m points while d of the m points lie on each of the n lines. The four numbers are not independent but evidently satisfy the equation

$$cm = dn.$$

The dual of (m_c, n_d) is (n_d, m_c). For instance, $(4_3, 6_2)$ is a quadrangle and $(6_2, 4_3)$ is a quadrilateral.

In the case of a *self-dual* configuration, we have $m = n$, $c = d$, and the symbol (n_d, n_d) is conveniently abbreviated to n_d. For instance, 3_2 is a triangle. We see from Figure 2.3A that 2.32 (Desargues's theorem) establishes the existence of a self-dual configuration 10_3: ten points and ten lines, with three points on each line and three lines through each point. In fact, the ten points

$$P, Q, R, P', Q', R', D, E, F, O$$

lie by threes on ten lines, as follows:

$$DQ'R', ER'P', FP'Q', DQR, ERP, FPQ, OPP', OQQ', ORR', DEF.$$

Referring to the three-dimensional proof of 2.31 (in the middle of page 19), we see that these ten points lie respectively on the ten joins of the five "exterior" points P_1, Q_1, R_1, S, S':

$$P_1S, Q_1S, R_1S, P_1S', Q_1S', R_1S', Q_1R_1, R_1P_1, P_1Q_1, SS',$$

while the ten lines lie respectively in the ten planes through triads of these same five points. Associating the five points P_1, Q_1, R_1, S, S' with the digits 1, 2, 3, 4, 5, we thus obtain a symmetric notation in which the points and lines of the Desargues configuration 10_3 are (in the above order):

$$G_{14}, G_{24}, G_{34}, G_{15}, G_{25}, G_{35}, G_{23}, G_{13}, G_{12}, G_{45},$$

$$g_{14}, g_{24}, g_{34}, g_{15}, g_{25}, g_{35}, g_{23}, g_{13}, g_{12}, g_{45}.$$

Whenever $i < j$, g_{ij} is the line containing three points whose subscripts involve neither i nor j, and G_{ij} is the common point of three such lines. In other words, a point and line of the configuration are incident if and only if their four subscripts are all different.

EXERCISES

1. Copy Figure 2.3A, marking the lines with the symbols g_{ij}.

2. Draw the five lines g_{12}, g_{23}, g_{34}, g_{45}, g_{15} in one color, and the five lines g_{14}, g_{24}, g_{25}, g_{35}, g_{13} in another color, thus exhibiting the configuration as a pair of simple pentagons, "mutually inscribed" in the sense that consecutive sides of each pass through alternate vertices of the other. (J. T. Graves, *Philosophical Magazine* (3), **15** (1839), p. 132.)

3. Beginning afresh, draw the four lines g_{15}, g_{25}, g_{35}, g_{45} in one color, and the remaining six in another color, thus exhibiting the configuration as a complete quadrilateral and a complete quadrangle, so situated that each vertex of the former lies on a side of the latter.

4. Is it possible to draw a configuration 7_3? [*Hint:* Look at Axiom 2.17.]

3.3 The Invariance of the Harmonic Relation

Dualizing Figure 2.5A, we find that any three concurrent lines a, b, c determine a fourth line f, concurrent with them, which we call the *harmonic conjugate of* c with respect to a and b. To construct it, we draw a triangle

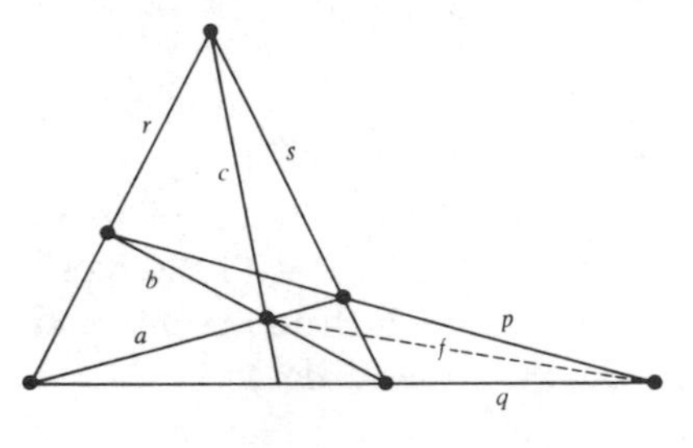

FIGURE 3.3A

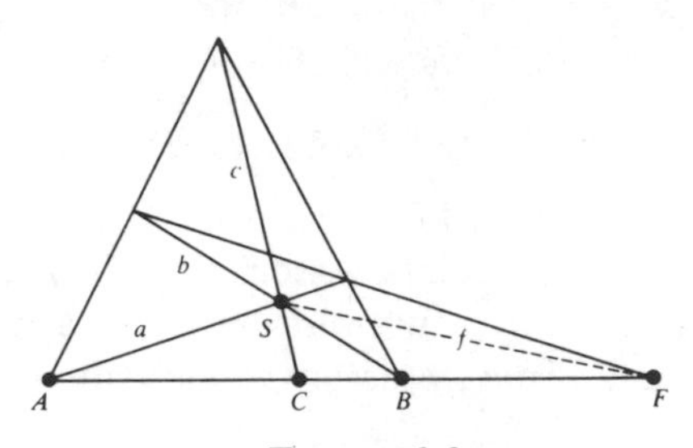

FIGURE 3.3B

qrs (Figure 3.3A) whose vertices $q \cdot r$, $q \cdot s$, $r \cdot s$ lie on a, b, c, respectively; then

$$p = (a \cdot s)(b \cdot r), \qquad f = (a \cdot b)(p \cdot q).$$

In fact, the quadrilateral $pqrs$ has a and b for two of its three diagonal lines while c and f, respectively, pass through the vertices that would be joined by the third diagonal line.

Figure 3.3B, which is obtained by identifying the lines

$$p,\ q,\ r,\ s,\ a,\ b,\ c$$

of Figure 3.3A with the lines

$$PQ,\ AB,\ QR,\ RP,\ PS,\ QS,\ RS$$

of Figure 2.5A, shows how the harmonic set of points $ABCF$ arises as a section of the harmonic set of lines $abcf$. Since such a figure can be derived from any harmonic set of points and any point S outside their line,

3.31 *A harmonic set of points is projected from any point outside the line by a harmonic set of lines.*

Dually,

3.32 *Any section of a harmonic set of lines, by a line not passing through the point of concurrence, is a harmonic set of points.*

Combining these two dual statements, we deduce that

3.33 *If* $ABCF \;\overline{\wedge}\; A'B'C'F'$ *and* $\mathsf{H}(AB, CF)$, *then* $\mathsf{H}(A'B', C'F')$.

In other words, perspectivities preserve the harmonic relation. By repeated application of this principle, we deduce:

3.34 *If* $ABCF \barwedge A'B'C'F'$ *and* $\mathsf{H}(AB, CF)$, *then* $\mathsf{H}(A'B', C'F')$.

In other words, projectivities preserve the harmonic relation. (In von Staudt's treatment, this property serves to *define* a projectivity. See Reference 7, p. 42.)

By Theorem 1.63 in the form $ABCF \barwedge FCBA$, we can now assert:

3.35 *If* $\mathsf{H}(AB, CF)$ *then* $\mathsf{H}(FC, BA)$,

and therefore also $\mathsf{H}(CF, BA)$, $\mathsf{H}(CF, AB)$, $\mathsf{H}(FC, AB)$.

EXERCISES

1. Let ABC be the diagonal triangle of a quadrangle $PQRS$. How can PQR be reconstructed if only ABC and S are given? [*Hint:* QR is the harmonic conjugate of AS with respect to AB and AC.]

2. If PQR is a triangle and $\mathsf{H}(AA_1, QR)$ and $\mathsf{H}(BB_1, RP)$, then P and Q are harmonic conjugates with respect to

$$C = AB_1 \cdot BA_1 \quad \text{and} \quad C_1 = AB \cdot A_1B_1.$$

3.4 Trilinear Polarity

It is interesting to see how Poncelet used a triangle to induce a correspondence between points not on its sides and lines not through its vertices. Let PQR be the triangle, and S a point of general position. The *Cevians* SP, SQ, SR determine points A, B, C on the sides QR, RP, PQ, as in Figure 3.4A. Let D, E, F be the harmonic conjugates of these points A, B, C, so that

$$\mathsf{H}(QR, AD), \ \mathsf{H}(RP, BE), \ \mathsf{H}(PQ, CF).$$

Comparing Figure 3.4A with Figure 2.5A, we see that these points D, E, F are the intersections of pairs of corresponding sides of the two triangles PQR and ABC, namely,

$$D = QR \cdot BC, \ E = RP \cdot CA, \ F = PQ \cdot AB.$$

Since these two triangles are perspective from S, 2.32 tells us that D, E, F lie on a line s. This line is what Poncelet called the *trilinear polar of* S.

Conversely, if we are given the triangle PQR and any line s, not through a vertex, we can define A, B, C to be the harmonic conjugates of the points D, E, F in which s meets the sides QR, RP, PQ. By 2.31, the three lines PA, QB, RC all pass through a point S, which is the *trilinear pole* of s.

EXERCISES

1. What happens if we stretch the definitions of trilinear pole and trilinear polar to the forbidden positions of S and s (namely S on a side, or s through a vertex)?

2. Locate (in Figure 3.4A) the trilinear pole of s with respect to the triangle ABC.

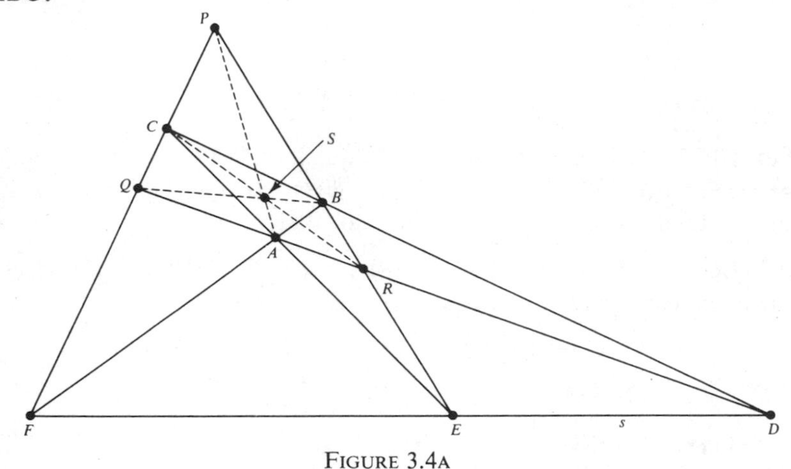

FIGURE 3.4A

3. How should all the lines in this figure be named so as to make it obviously self-dual?

4. In the spirit of Section 2.5, Exercise 5, what metrical property of the triangle PQR arises when we seek the trilinear pole of the line at infinity?

3.5 Harmonic Nets

A point P is said to be *harmonically related* to three distinct collinear points A, B, C if P can be exhibited as a member of a sequence of points beginning with A, B, C and proceeding according to this rule: each point (after C) forms a harmonic set with three previous points. (Any three previous points can be used, in any order.) The set of all points harmonically related to A, B, C is called a *harmonic net* or "net of rationality" (see Reference **19**, p. 84), and is denoted by R(ABC). Of course, the same harmonic net is equally well denoted by R(BAC) or R(BCA), and so forth. Since a projectivity transforms any harmonic set into a harmonic set, it also transforms any harmonic net into a harmonic net. In particular,

3.51 *If a projectivity leaves invariant each of three distinct points A, B, C on a line, it leaves invariant every point of the harmonic net* $\mathsf{R}(ABC)$.

This result, which we have deduced from our first seven axioms, may reasonably be regarded as making Axiom 2.18 plausible (see page 16).

By Theorem 1.63 (with B and D interchanged), any four collinear points A, B, C, D satisfy $ABCD \barwedge DCBA$. If D belongs to $\mathsf{R}(ABC)$, A must belong to $\mathsf{R}(DCB)$, which is the same as $\mathsf{R}(BCD)$. It follows that not only A, but every point of $\mathsf{R}(ABC)$, belongs also to $\mathsf{R}(BCD)$. Hence, if D belongs to $\mathsf{R}(ABC)$,

$$\mathsf{R}(ABC) = \mathsf{R}(BCD).$$

By repeated applications of this result we see that, if K, L, M are any three distinct points of $\mathsf{R}(ABC)$,

$$\mathsf{R}(ABC) = \mathsf{R}(BCK) = \mathsf{R}(CKL) = \mathsf{R}(KLM).$$

Hence

3.52 *A harmonic net is equally well determined by any three distinct points belonging to it.*

We could, in fact, define a harmonic net to be a set, as small as possible, of at least three collinear points which includes, with every three of its members, the harmonic conjugate of each with respect to the other two.

If we begin to make a careful drawing of a harmonic net, we soon find that this is an endless task: the harmonic net seems to include infinitely many points between any two given points, thus raising the important question whether it includes every point on the line. There is some advantage in leaving this question open, that is, answering "Yes and No." Projective geometry, as we are developing it here, is not *categorical:* it is really not just one geometry but many geometries, depending on the nature of the set of all points on a line. In *rational* geometry and the simplest *finite* geometries, all the points on a line form a single harmonic net, so that Axiom 2.18 becomes superfluous. In *real* geometry, on the other hand, the points on a line are arranged like the continuum of real numbers, among which the rational numbers represent a typical harmonic net (Reference 7, p. 179). Between any two rational numbers, no matter how close, we can find infinitely many other rational numbers and also infinitely many irrational numbers. Analogously, between any two points of a harmonic net on the *real* line we can find infinitely many other members of the net and also infinitely many points *not* belonging to the harmonic net. Axiom 2.18 (or, alternatively, some statement about continuity) is needed to ensure the invariance of these extra points.

When the points of a line in rational geometry, or the points of a harmonic

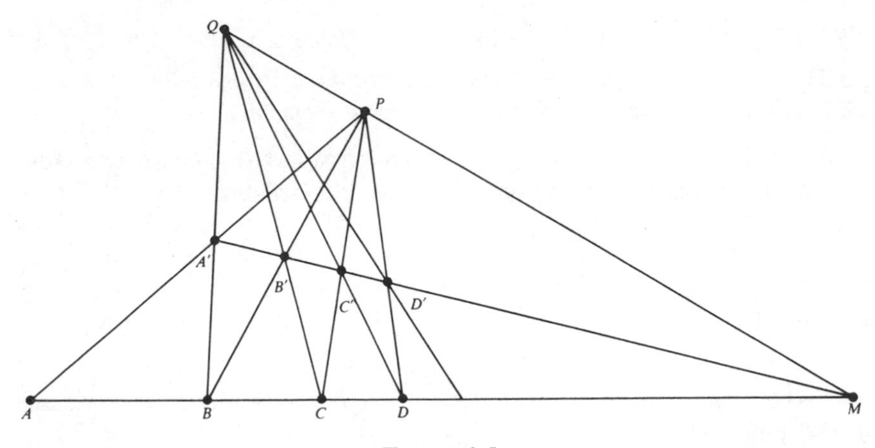

FIGURE 3.5A

net in real geometry, are represented by the rational numbers, the *natural* numbers (that is, positive integers) represent a *harmonic sequence* $ABCD\ldots$, which is derived from three collinear points A, B, M by the following special procedure. With the help of two arbitrary points P and Q on another line through M, as in Figure 3.5A, we construct in turn,

$$A' = AP \cdot BQ, \qquad B' = BP \cdot A'M,$$

$$C = B'Q \cdot AM, \qquad C' = CP \cdot A'M,$$

$$D = C'Q \cdot AM, \qquad D' = DP \cdot A'M,$$

and so on. In view of the harmonic relations

$$\mathsf{H}(BM, AC), \ \mathsf{H}(CM, BD), \ldots,$$

the sequence $ABC\ldots$, so constructed, depends only on the given points A, B, M, and is independent of our choice of the auxiliary points P and Q.

EXERCISES

1. Is the harmonic sequence $ABC\ldots$ uniquely determined by its first three members?

2. In the notation of Figure 3.5A, is $A'B'C'\ldots$ a harmonic sequence?

3. What happens to the sequence $ABCD\ldots$ (Figure 3.5A) when PQ is the line at infinity, so that $ABB'A'$, $BCC'B'$, $CDD'C'$, ..., are parallelograms?

CHAPTER FOUR

The Fundamental Theorem and Pappus's Theorem

> The Golden Age of Greek geometry ended with the time of Apollonius of Perga. . . . The beginning of the Christian era sees quite a different state of things. . . . Production was limited to elementary textbooks of decidedly feeble quality. . . . The study of higher geometry languished or was completely in abeyance until Pappus arose to revive interest in the subject. . . . The great task which he set himself was the re-establishment of geometry on its former high plane of achievement.
>
> *Sir Thomas Heath* (1861–1940)
> (Reference **11**, p. 355)

4.1 How Three Pairs Determine a Projectivity

Given four distinct points A, B, C, X on one line, and three distinct points A', B', C' on the same or another line, there are many possible ways in which we may proceed to construct a point X' (on $A'B'$) such that

$$ABCX \;\overline{\wedge}\; A'B'C'X'.$$

For instance, if the two lines are distinct, one way is indicated in Figure 4.1A (which is Figure 1.6B in a revised notation):

(4.11) $$ABCX \overset{A'}{\overline{\overline{\wedge}}} GNMQ \overset{A}{\overline{\overline{\wedge}}} A'B'C'X'.$$

This can be varied by using B' and B (or C' and C), instead of A' and A, as centers of the two perspectivities. If, on the other hand, the given points are all on one line, as in Figure 4.1B, we can use an arbitrary perspectivity $ABCX \doublebarwedge A_1B_1C_1X_1$ to obtain four points on another line, and then relate $A_1B_1C_1X_1$ to $A'B'C'X'$, so that altogether

$$ABCX \overset{O}{\doublebarwedge} A_1B_1C_1X_1 \overset{A'}{\doublebarwedge} GNMQ \overset{A_1}{\doublebarwedge} A'B'C'X'.$$

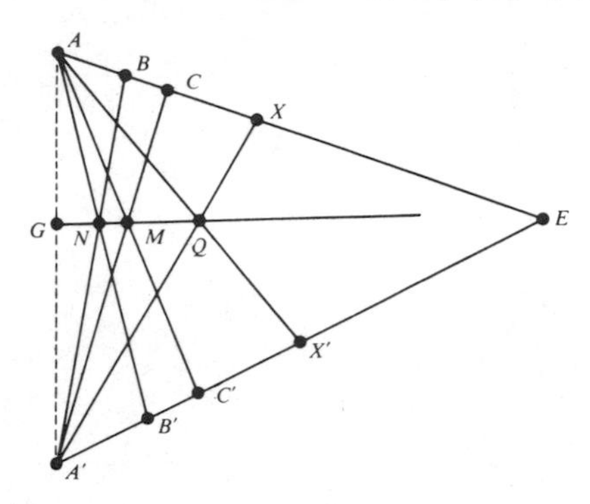

FIGURE 4.1A

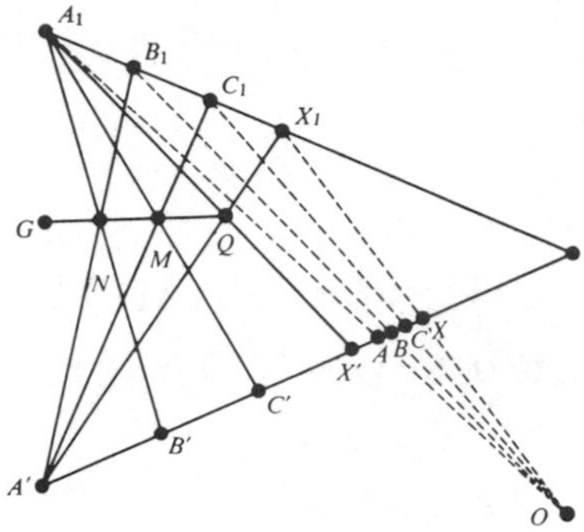

FIGURE 4.1B

We saw, in Figure 2.4B, that five collinear points A, B, C, D, E determine a unique sixth point F such that $(AD)(BE)(CF)$. By declaring F to be unique, we mean that its position is independent of our choice of the auxiliary triangle QRS. Can we say, analogously, that seven points A, B, C, X, A', B', C' (with the first four, and likewise the last three, collinear and distinct) determine a unique eighth point X' such that

$$ABCX \barwedge A'B'C'X'?$$

If not, there must be two distinct chains of perspectivities yielding, respectively,

$$ABCX \barwedge A'B'C'X' \quad \text{and} \quad ABCX \barwedge A'B'C'X'',$$

where $X'' \neq X'$. Proceeding backwards along the first chain and then forwards along the second, we would have

$$A'B'C'X' \barwedge A'B'C'X'',$$

contradicting Axiom 2.18. Thus, by *reductio ad absurdum*, we have proved

4.12 THE FUNDAMENTAL THEOREM OF PROJECTIVE GEOMETRY: *A projectivity is determined when three collinear points and the corresponding three collinear points are given.*

Of course, either of the sets of "three collinear points" (or both) can be replaced by "three concurrent lines." Thus each of the relations

$$ABC \barwedge A'B'C', \quad ABC \barwedge abc, \quad abc \barwedge ABC, \quad abc \barwedge a'b'c'$$

suffices to specify uniquely a particular projectivity. On the other hand, each of the relations

$$ABCD \barwedge A'B'C'D', \quad ABCD \barwedge abcd, \quad abcd \barwedge a'b'c'd'$$

expresses a special property of eight points, or of four points and four lines, or of eight lines, of such a nature that any seven of the eight will uniquely determine the remaining one. We now have the proper background for Theorem 1.63, which tells us that, if a projectivity interchanges A and B while transforming C into D, it also transforms D into C, that is, it interchanges C and D.

EXERCISES

1. Given three collinear points A, B, C, set up three perspectivities whose product has the effect $ABC \barwedge BCA$.
2. If the projectivity $ABC \barwedge BCA$ transforms D into E, and E into F, how does it affect F? [*Hint:* Use Axiom 2.18.]
3. If A, B, C, D are distinct collinear points, then

$$ABCD \barwedge BADC \barwedge CDAB \barwedge DCBA.$$

4.2 Some Special Projectivities

The following simple consequences of 4.12 will be found useful.

4.21 *Any two harmonic sets (of four collinear points or four concurrent lines) are related by a unique projectivity.*

PROOF. If $\mathsf{H}(AB, CF)$ and $\mathsf{H}(A'B', C'F')$, the projectivity $ABC \barwedge A'B'C'$ transforms F into a point F'' such that, by 3.34, $\mathsf{H}(A'B', C'F'')$. But the harmonic conjugate of C' with respect to A' and B' is unique. Therefore F'' coincides with F'. The same reasoning can be used when either of the harmonic sets consists of lines instead of points.

4.22 *A projectivity relating ranges on two distinct lines is a perspectivity if and only if the common point of the two lines is invariant.*

PROOF. A perspectivity obviously leaves invariant the common point of the two lines. On the other hand, if a projectivity relating ranges on two distinct lines has an invariant point E, this point, belonging to both ranges, must be the common point of the two lines, as in Figure 4.2A. Let A and B

be two other points of the first range, A' and B' the corresponding points of the second. The fundamental theorem tells us that the perspectivity

$$ABE \overset{O}{\barwedge} A'B'E,$$

where $O = AA' \cdot BB'$, is the same as the given projectivity $ABE \barwedge A'B'E$.

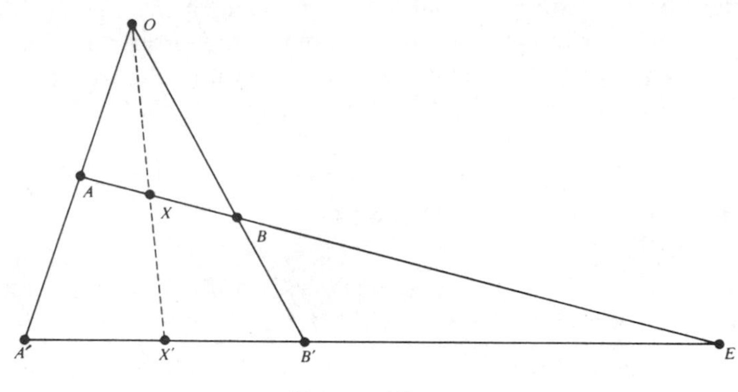

FIGURE 4.2A

EXERCISES

1. Any two harmonic nets, R(ABC) and R($A'B'C'$) (or any two harmonic sequences, $ABC\ldots$ and $A'B'C'\ldots$) are related by a projectivity.
2. Let the side QR of a quadrangle $PQRS$ meet the side BC of its diagonal triangle ABC in A_1, as in Exercise 2 of Section 3.1. Regarding $ABPCSA_1$ as a hexagon whose six vertices lie alternately on two lines, what can be said about the intersections of pairs of "opposite" sides of this hexagon?
3. Dualize Theorem 4.22 and Figure 4.2A.
4. If H(AB, CD) then $ABCD \barwedge BACD$. (Compare Section 4.1, Exercise 3.)

4.3 The Axis of a Projectivity

The fundamental theorem 4.12 tells us that there is only one projectivity $ABC \barwedge A'B'C'$ relating three distinct points on one line to three distinct points on the same or any other line. The construction indicated in 4.11 (see Figure 4.1A) shows how, when the lines AB and $A'B'$ are distinct, this unique projectivity can be expressed as the product of two perspectivities whose centers may be any pair of corresponding points (in reversed order) of the two related ranges. It is natural to ask whether different choices of the two centers will yield different positions for the line MN of the intermediate range. A negative answer is provided by the following theorem:

4.31 *Every projectivity relating ranges on two distinct lines determines another special line, the "axis," which contains the intersection of the cross-joins of any two pairs of corresponding points.*

PROOF. In the notation of 4.11, the perspectivities from A' and A determine the axis MN, which contains the common point of the "cross-joins" of the pair AA' and any other pair; for instance, N is the common point of the cross-joins (AB' and BA') of the two pairs AA' and BB'. What remains to be proved is the *uniqueness* of this axis: that another choice of the two pairs of corresponding points (such as BB' and CC') will yield another "crossing point" on the same axis. For this purpose, we seek a new specification for the axis, independent of the particular pair AA'.

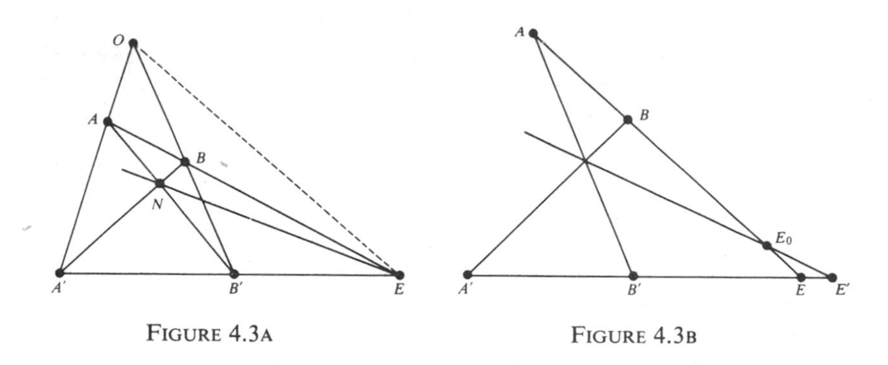

FIGURE 4.3A

FIGURE 4.3B

Let E be the common point of the two lines. Suppose first that E is invariant, as in Figures 4.2A and 4.3A. Referring to Figure 4.1A, we observe that, when X coincides with E, so also do Q and X'. The axis EN is independent of AA', since it can be described as *the harmonic conjugate of EO with respect to the given lines EB and EB'*. On the other hand, if the common point E is noninvariant, as in Figure 4.3B, it is still a point belonging to both ranges. Referring to Figure 4.1A again, we observe that, when X coincides with E, Q and X' both coincide with the corresponding point E'. Hence the axis passes through E'. For a similar reason the axis also passes through the point E_0 of the first range for which the corresponding point of the second is E. Hence the axis can be described as *the join* E_0E'.

This completes the proof of 4.31. The final remark can be expressed as follows:

4.32 *If E_0EE' is a triangle, the axis of the projectivity $AE_0E \barwedge A'EE'$ is the line E_0E'.*

EXERCISE

Dualize Theorem 4.31 and Figure 4.3B.

4.4 Pappus and Desargues

We are now ready to prove one of the oldest of all projective theorems. Pappus of Alexandria, living in the fourth century A.D., used a laborious development of Euclid's methods (see Reference **21**, pp. 286–290, and **8a**, pp. 66–69).

4.41 PAPPUS'S THEOREM: *If the six vertices of a hexagon lie alternately on two lines, the three pairs of opposite sides meet in collinear points.*

PROOF. Let the hexagon be $AB'CA'BC'$, as Figure 4.4A (which shows two of the many different ways in which it can be drawn). Since alternate vertices are collinear, there is a projectivity $ABC \barwedge A'B'C'$. The pairs of opposite sides, namely

$$B'C, BC'; \qquad C'A, CA'; \qquad A'B, AB',$$

are just the cross-joins of the pairs of corresponding points

$$BB', CC'; \qquad CC', AA'; \qquad AA', BB'.$$

By Theorem 4.31, their points of intersection

$$L = B'C \cdot BC', \qquad M = C'A \cdot CA', \qquad N = A'B \cdot AB'$$

all lie on the axis of the projectivity.

Alternatively, using further points

$$J = AB' \cdot CA', \qquad E = AB \cdot A'B', \qquad K = AC' \cdot CB',$$

we have

$$ANJB' \stackrel{A'}{\barwedge} ABCE \stackrel{C'}{\barwedge} KLCB'.$$

Thus B' is an invariant point of the projectivity $ANJ \barwedge KLC$. By Theorem 4.22, this projectivity is a perspectivity, namely

$$ANJ \stackrel{M}{\barwedge} KLC.$$

Thus M lies on NL; that is, L, M, N are collinear.

Similarly, since Theorem 4.22 is a consequence of the five axioms 2.13, 3.11, 3.12, 2.17, 2.18, we could have proved 2.32 (Desargues's theorem) as follows.

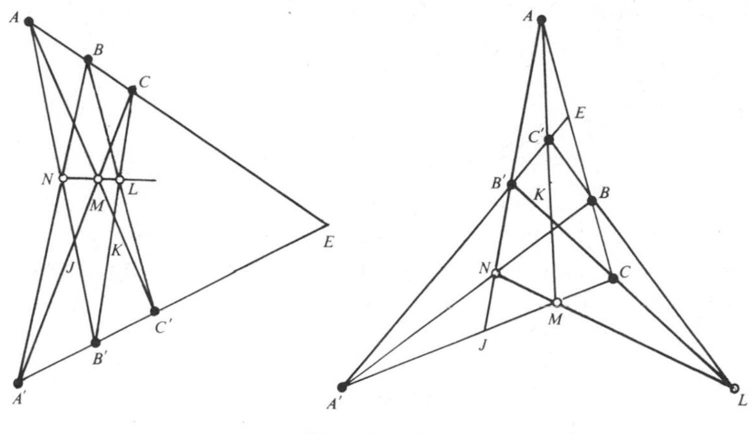

FIGURE 4.4A

Assuming that the lines PP', QQ', RR' all pass through O, as in Figure 2.3A, we define $D = QR \cdot Q'R'$, $E = RP \cdot R'P'$, $F = PQ \cdot P'Q'$ and three further points

$$A = OP \cdot DE, \qquad B = OQ \cdot DE, \qquad C = OR \cdot DE.$$

Then

$$OPAP' \overset{E}{\barwedge} ORCR' \overset{D}{\barwedge} OQBQ'.$$

so that O is an invariant point of the projectivity $PAP' \barwedge QBQ'$. By 4.22, this projectivity is a perspectivity. Its center, F, lies on AB, which is DE; that is, D, E, F are collinear.

This procedure (see Reference **20**, p. 16) has the advantage of allowing us to omit 2.32 from the list of two-dimensional axioms proposed in Section 3.1. (The list is thus seen to be redundant: only five of the six are really needed.) It has the disadvantage of making Theorem 3.51 depend on Axiom 2.18, so that the theorem can no longer be used to make the axiom plausible!

We saw, in Section 3.2, that the figure for Desargues's theorem is a self-dual configuration 10_3. Somewhat analogously, the figure for Pappus's theorem is a self-dual configuration 9_3: nine points and nine lines, with three points on each line and three lines through each point. This fact becomes evident as soon as we have made the notation more symmetrical by calling the nine points

$$A_1 = A,\ \ B_1 = B,\ \ C_1 = C,\ \ A_2 = A',\ \ B_2 = B',\ \ C_2 = C',$$
$$A_3 = L,\ \ B_3 = M,\ \ C_3 = N$$

and the nine lines

$$a_1 = BL,\quad b_1 = AM,\quad c_1 = A'B',\quad a_2 = CM,\quad b_2 = CL,\quad c_2 = AB,$$
$$a_3 = AN,\quad b_3 = BN,\quad c_3 = LM.$$

The three triangles

$$A_1B_1C_2,\, A_2B_2C_3,\, A_3B_3C_1 \quad \text{or} \quad a_1b_1c_2,\, a_3b_3c_1,\, a_2b_2c_3$$

provide an instance of *Graves triangles:* a cycle of three triangles, each inscribed in the next. (See page 130 of Graves's paper mentioned in Exercise 2 of Section 3.2. This aspect of the Pappus configuration was rediscovered by G. Hessenberg.)

EXERCISES

1. Using the "symmetrical" notation for the Pappus configuration, prepare a table showing which are the three points on each line and which are the three lines through each point. Observe that the three points A_i, B_j, C_k are collinear (and the three lines a_i, b_j, c_k are concurrent) whenever $i + j + k$ is a multiple of three. (Reference **12a**, p. 108.)

2. Given a triangle $A_1A_2A_3$ and two points B_1, B_2, locate a point B_3 such that the three lines A_1B_1, A_2B_3, A_3B_2 are concurrent while also the three lines A_1B_3, A_2B_2, A_3B_1 are concurrent. Then also the three lines A_1B_2, A_2B_1, A_3B_3 are concurrent; in other words, *if two triangles are doubly perspective, they are triply perspective.* (Reference **19**, p. 100.)

3. What can be said about the three "diagonals" of the hexagon $A_1B_3A_2B_2A_3B_1$?

4. State the dual of Pappus's theorem and name the sides of the hexagon in the notation of Exercise 1.

5. Consider your solution to Exercise 2 of Section 4.2. Could this be developed into a proof of Pappus's theorem?

6. If one triangle is inscribed in another, any point on a side of the former can be used as a vertex of a third triangle which completes a cycle of Graves triangles (each inscribed in the next).

7. Assign the digits 0, 1, . . . , 8 to the nine points of the Pappus configuration in such a way that 801, 234, 567 are three triads of collinear points while 012, 345, 678 is a cycle of Graves triangles. (E. S. Bainbridge.*)

---

* In his answer to an examination question.

CHAPTER FIVE

One-Dimensional Projectivities

The most original and profound of the projective geometers of the German school was Georg Karl Christian von Staudt, long professor at Erlangen. His great passion . . . was for unity of method.

J. L. Coolidge (1873–1954)
(Reference **4**, p. 61)

5.1 Superposed Ranges

Axiom 2.18 tells us that a projectivity relating two ranges on one line (that is, a projective transformation of the line into itself) cannot have more than two invariant points without being merely the *identity*, which relates each point to itself. The projectivity is said to be *elliptic*, *parabolic*, or *hyperbolic* according as the number of invariant points is 0, 1, or 2. We shall see that both parabolic and hyperbolic projectivities always exist. In fact, Figure 5.1A (which is essentially the same as Figure 2.4A) suggests a simple construction for the hyperbolic or parabolic projectivity $AEC \barwedge BDC$ which has a given invariant point C. Here A, B, C, D, E are any five collinear points, and we draw a quadrangle $PQRS$ as if we were looking for the sixth point of a quadrangular set. The given projectivity can be expressed as the product of two perspectivities

$$AEC \overset{P}{\barwedge} SRC \overset{Q}{\barwedge} BDC,$$

and it is easy to see what happens to any other point on the line.

Evidently C (on RS) is invariant. If any other point is invariant, it must be collinear with the centers P and Q of the two perspectivities; that is, it can only be F. Hence the projectivity $AEC \barwedge BDC$ is hyperbolic if C and F are distinct (Figure 5.1A) and parabolic if they are coincident (Figure 5.1B). In the former case we have $AECF \barwedge BDCF$, and in the latter we naturally write

$$AECC \barwedge BDCC,$$

the repeated letter indicating that the projectivity is parabolic. Thus

5.11 *The two statements $AECF \barwedge BDCF$ and $(AD)(BE)(CF)$ are equivalent, not only when C and F are distinct but also when they coincide.*

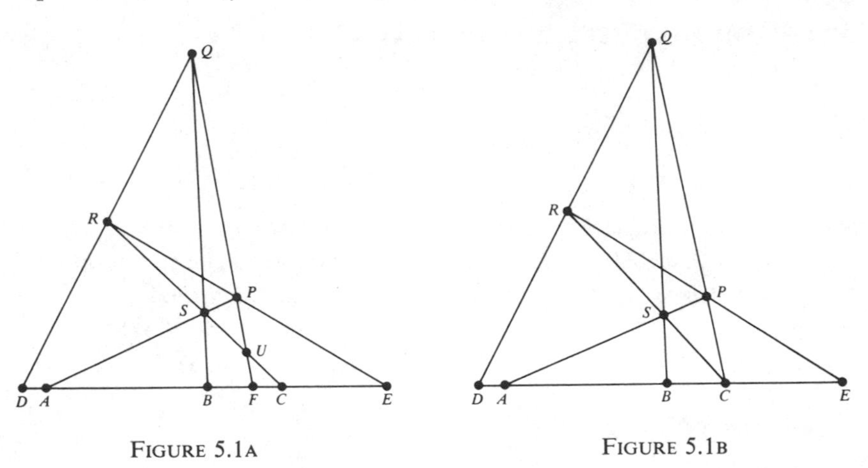

FIGURE 5.1A

FIGURE 5.1B

Since the statement $AECF \barwedge BDCF$ involves C and F symmetrically, the statement $(AD)(BE)(CF)$ is equivalent to $(AD)(BE)(FC)$, and similarly to $(AD)(EB)(FC)$ and to

$$(DA)(EB)(FC).$$

This is remarkable because, when the quadrangular set is derived from the quadrangle, the two triads ABC and DEF arise differently: the first from three sides with a common vertex, and the second from three that form a triangle. It is interesting that, whereas one way of matching two quadrangles (Figure 2.4B) uses only Desargues's theorem, the other (Exercise 2, below) needs the fundamental theorem.

Our use of the words *elliptic*, *parabolic* and *hyperbolic* arises from the aspect of the projective plane as a Euclidean plane with an extra line at infinity. In this aspect a conic is an ellipse, a parabola, or a hyperbola according as its number of points at infinity is 0, 1, or 2, that is, according as it "goes off to infinity" not at all, or in one direction (the direction of the axis of the parabola) or in two directions (the directions of the asymptotes of the hyperbola). The names of the conics themselves are due to Apollonius (see Reference **13**, p. 10).

EXERCISES

1. Using Figure 5.1A again (and defining $V = PS \cdot QR$), construct a hyperbolic projectivity having A and D as its invariant points.
2. Take any five collinear points A, B, C, D, E. Construct F so that $(AD)(BE)(CF)$. Then (on the other side of the line, for convenience) construct C' so that $(DA)(EB)(FC')$. See how nearly your C' agrees with C.
3. Two perspectivities cannot suffice for the construction of an elliptic projectivity. In other words, if an elliptic projectivity exists, its construction requires three perspectivities.
4. Does an elliptic projectivity exist?

5.2 Parabolic Projectivities

The fundamental theorem, 4.12, shows that a hyperbolic projectivity is determined when both its invariant points and one pair of distinct corresponding points are given. In fact, any four collinear points A, B, C, F determine such a projectivity $ACF \barwedge BCF$, with invariant points C and F. To construct it, we take a triangle QPS whose sides PS, SQ, QP pass, respectively, through A, B, F. If the side through F meets CS in U (as in Figure 5.1A), we have

$$ACF \overset{P}{\barwedge} SCU \overset{Q}{\barwedge} BCF.$$

If we regard E as an arbitrary point on the same line AB, this construction yields the corresponding point D. It remains effective when C, F and U coincide (as in Figure 5.1B), that is, when the line AB passes through the diagonal point $U = PQ \cdot RS$ of the quadrangle. Thus a parabolic projectivity is determined when its single invariant point and one pair of distinct corresponding points are given. We naturally call it *the projectivity* $ACC \barwedge BCC$. This notation is justified by its transitivity:

5.21 *The product of two parabolic projectivities having the same invariant point is another such parabolic projectivity (if it is not merely the identity).*

PROOF. Clearly, the common invariant point C of the two projectivities is still invariant for the product, which is therefore either parabolic or hyperbolic. The latter possibility is excluded by the following argument. If any other point A were invariant for the product, the first parabolic projectivity would take A to some different point B, and the second would take B back to A. Thus the first would be $ACC \barwedge BCC$, the second would be its inverse $BCC \barwedge ACC$, and the product would not be properly hyperbolic but merely the identity.

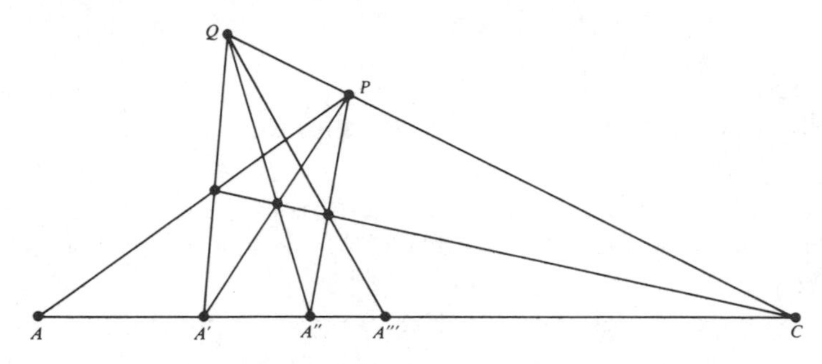

FIGURE 5.2A

Thus the product of $ACC \barwedge A'CC$ and $A'CC \barwedge A''CC$ is $ACC \barwedge A''CC$, and we can safely write out strings of parabolic relations such as

$$ABCC \barwedge A'B'CC \barwedge A''B''CC.$$

In particular, by "iterating" a parabolic projectivity $ACC \barwedge A'CC$ we obtain a sequence of points $A, A', A'', \ldots$ such that

$$CCAA'A'' \ldots \barwedge CCA'A''A''' \ldots,$$

as in Figure 5.2A. Comparing this with Figure 3.5A, we see that $AA'A'' \ldots$ is a harmonic sequence.

We have seen that the statements

$$AECF \barwedge BDCF \quad \text{and} \quad (AD)(BE)(CF)$$

are equivalent. Setting $B = E$ and $C = F$, we deduce the equivalence of

$$ABCC \barwedge BDCC \quad \text{and} \quad \mathsf{H}(BC, AD).$$

Hence, after a slight change of notation,

5.22 *The projectivity* $AA'C \barwedge A'A''C$ *is parabolic if* $\mathsf{H}(A'C, AA'')$, *and hyperbolic otherwise.*

In other words, the parabolic projectivity $ACC \barwedge A'CC$ transforms A' into the harmonic conjugate of A with respect to A' and C.

EXERCISE

What happens to the parabolic projectivity $ACC \barwedge A'CC$ (Figure 5.2A) when PQ is the line at infinity (as in Exercise 3 of Section 3.5)?

5.3 Involutions

Desargues's works were not well received during his lifetime. This lack of appreciation was possibly a result of his obscure style; he introduced about seventy new terms, of which only *involution* has survived. According to his definition, formulated in terms of the nonprojective concept of distance and the arithmetic concept of multiplication, an "involution" is the relation between pairs of points on a line whose distances from a fixed point have a constant product (positive or negative). He might well have added "or a constant sum." An equivalent definition, not using distances, was given by von Staudt: *An involution is a projectivity of period two*, that is, a projectivity which *interchanges* pairs of points. It is remarkable that this relation

$$XX' \barwedge X'X$$

holds for all positions of X if it holds for any one position:

5.31 *Any projectivity that interchanges two distinct points is an involution.*

PROOF. Let $X \barwedge X'$ be the given projectivity which interchanges two distinct points A and A', so that

$$AA'X \barwedge A'AX',$$

where X is an arbitrary point on the line AA'. By Theorem 1.63, there is a projectivity in which $AA'XX' \barwedge A'AX'X$. By the fundamental theorem 4.12, this projectivity, which interchanges X and X', is the same as the given projectivity. Since X was arbitrarily chosen, the given projectivity is an involution.

Any four collinear points A, A', B, B' determine a projectivity $AA'B \barwedge A'AB'$, which we now know to be an involution. Hence

5.32 *An involution is determined by any two of its pairs.*

Accordingly, it is convenient to denote the involution $AA'B \barwedge A'AB'$ by

$$(AA')(BB')$$

or $(A'A)(BB')$, or $(BB')(AA')$, and so forth. This notation remains valid when B' coincides with B; in other words, the involution $AA'B \barwedge A'AB$, for which B is invariant, may be denoted by

$$(AA')(BB).$$

If $(AD)(BE)(CF)$, as in Figure 2.4A, we can combine the projectivity $AECF \barwedge BDCF$ of 5.11 with the involution $(BD)(CF)$ to obtain

$$AECF \barwedge BDCF \barwedge DBFC,$$

which shows that there is a projectivity in which $AECF \barwedge DBFC$. Since this interchanges C and F, it is an involution, namely

$$(BE)(CF) \quad \text{or} \quad (CF)(AD) \quad \text{or} \quad (AD)(BE).$$

Thus the quadrangular relation $(AD)(BE)(CF)$ is equivalent to the statement that the projectivity $ABC \barwedge DEF$ is an involution, or that

$$ABCDEF \barwedge DEFABC.$$

In other words,

5.33 *The three pairs of opposite sides of a complete quadrangle meet any line (not through a vertex) in three pairs of an involution. Conversely, any three collinear points, along with their mates in an involution, form a quadrangular set.*

It follows that the construction for F, when A, B, C, D, E are given (as in the preamble to 2.41), may be regarded as a construction for the mate of C in the involution $(AD)(BE)$. (See Figure 2.4A or 5.1A.)

We have seen that CF is a pair of the involution $(AD)(BE)$ if and only if $AECF \barwedge BDCF$. We must get accustomed to using other letters in the same context. For instance, *MN is a pair of the involution $(AB')(BA')$ if and only if $AA'MN \barwedge BB'MN$*. Since $(AB')(BA')$ is the same as $(AB')(A'B)$, it follows that the two statements

$$AA'MN \barwedge BB'MN \quad \text{and} \quad ABMN \barwedge A'B'MN$$

are equivalent. (Notice that it is only the *statements* that are equivalent: the two *projectivities* are, of course, distinct.)

If two involutions, $(AA')(BB')$ and $(AA_1)(BB_1)$, have a common pair MN, we deduce

$$A'B'MN \barwedge BAMN \barwedge A_1B_1MN.$$

Hence

5.34 *If MN is a pair of each of the involutions $(AA')(BB')$ and $(AA_1)(BB_1)$, it is also a pair of $(A'B_1)(B'A_1)$.*

All these results remain valid when M and N coincide, so that we are dealing with parabolic (instead of hyperbolic) projectivities. Thus M is an invariant point of the involution $(AB')(BA')$ if and only if $AA'MM \barwedge BB'MM$, that is, if and only if $ABMM \barwedge A'B'MM$; and if M is an invariant point of each of the involutions $(AA')(BB')$ and $(AA_1)(BB_1)$, it is also an invariant point of $(A'B_1)(B'A_1)$.

If two involutions have a common pair MN, their product is evidently hyperbolic, with invariant points M and N. In fact, by watching their effect on A, M, N in turn, we see that the product of $(AB)(MN)$ and $(BC)(MN)$ is $AMN \barwedge CMN$. More interestingly,

5.35 *Any one-dimensional projectivity is expressible as the product of two involutions.*

PROOF. Let the given projectivity be $ABC \barwedge A'B'C'$, where neither A nor B is invariant. By watching what happens to A, B, C in turn, we see that this projectivity has the same effect as the product of the two involutions

$$(AB')(BA') \quad \text{and} \quad (A'B')(C'D),$$

where D is the mate of C in $(AB')(BA')$. (J. L. Coolidge, *A Treatise on the Circle and the Sphere*, Clarendon Press, Oxford, 1916, p. 200.)

EXERCISES

1. Given six collinear points A, B, C, D, E, F, consider the three involutions $(AB)(DE)$, $(BC)(EF)$, $(CD)(FA)$. If any two of these involutions have a common pair, all three have a common pair.
2. The hyperbolic projectivity $MNA \barwedge MNA'$ is the product of the involutions $(AB)(MN)$ and $(A'B)(MN)$, where B is an arbitrary point on the line.
3. Any involution $(AA')(BB')$ may be expressed as the product of $(AB)(A'B')$ and $(AB')(BA')$.
4. Any projectivity that is not an involution may be expressed as the product of $(AA'')(A'A')$ and $(AA''')(A'A'')$.
5. Notice that Exercises 3 and 4 together provide an alternative proof for Theorem 5.35. In the proof given in the text, why was it necessary to insist that neither A nor B is invariant?

5.4 Hyperbolic Involutions

As we have seen, any involution that has an invariant point B (and a pair of distinct corresponding points C and C') may be expressed as $BCC' \barwedge BC'C$ or $(BB)(CC')$. Let A denote the harmonic conjugate of B with respect to C and C'. By 4.21, the two harmonic sets $ABCC'$ and $ABC'C$ are related by a unique projectivity $ABCC' \barwedge ABC'C$. The fundamental theorem identifies this with the given involution. Hence

5.41 *Any involution that has an invariant point B has another invariant point A, which is the harmonic conjugate of B with respect to any pair of distinct corresponding points.*

Thus any involution that is not elliptic is hyperbolic: there are no "parabolic involutions." Moreover, any two distinct points A and B are the invariant points of a unique hyperbolic involution, which is simply the correspondence between harmonic conjugates with respect to A and B. This is naturally denoted by

$$(AA)(BB).$$

The harmonic conjugate of C with respect to any two distinct points A and B may now be redefined as the mate of C in the involution $(AA)(BB)$. Unlike the definition of harmonic conjugate in Section 2.5, this new definition remains meaningful when C coincides with A or B:

5.42 *Any point is its own harmonic conjugate with respect to itself and any other point.*

EXERCISES

1. If $ABCD \barwedge BACD$ then $\mathsf{H}(AB, CD)$. (Compare Section 4.2, Exercise 4.)
2. If a hyperbolic projectivity relates two points that are harmonic conjugates with respect to the invariant points, it must be an involution.
3. If $\mathsf{H}(AB, MN)$ and $\mathsf{H}(A'B', MN)$, then MN is a pair of the involution $(AA')(BB')$. [*Hint:* $AB'MN \barwedge BA'MN$.]
4. Given $(AD)(BE)(CF)$, let A', B', C', D', E', F' be the harmonic conjugates of A, B, C, D, E, F with respect to two fixed points on the same line. Then $(A'D')(B'E')(C'F')$.
5. If $ABCD \barwedge ABDE$ and $\mathsf{H}(CE, DD')$, then $\mathsf{H}(AB, DD')$. (S. Schuster.)
6. Let X be a variable point collinear with three distinct points A, B, C, and let Y and X' be defined by $\mathsf{H}(AB, XY)$, $\mathsf{H}(BC, YX')$. Then the projectivity $X \barwedge X'$ is parabolic. [*Hint:* When X is invariant, so that X' coincides with it, Y too must coincide with it; for otherwise both A and C would be the harmonic conjugate of B with respect to X and Y. Therefore X must be either A or B, and also either B or C; that is, $X = B$.]
7. If $\mathsf{H}(BC, AD)$ and $\mathsf{H}(CA, BE)$ and $\mathsf{H}(AB, CF)$, then $(AD)(BE)(CF)$. [*Hint:* Apply 5.41 to the involution $BCAD \barwedge ACBE$. Deduce $\mathsf{H}(DE, CF)$.]

CHAPTER SIX

Two-Dimensional Projectivities

History shows that those heads of empires who have encouraged the cultivation of mathematics . . . are also those whose reigns have been the most brilliant and whose glory is the most durable.

Michel Chasles (1793–1880)
(Reference 3, p. 163)

6.1 Projective Collineations

We shall find that the one-dimensional projectivity $ABC \barwedge A'B'C'$ has two different analogues in two dimensions: one relating points to points and lines to lines, the other relating points to lines and lines to points. The names *collineation* and *correlation* were introduced by Möbius in 1827, but some special collineations (such as translations, rotations, reflections, and dilatations) were considered much earlier. Another example is Poncelet's "homology": the relation between the central projections of a plane figure onto another plane from two different centers of perspective. We shall give (in Section 6.2) a two-dimensional treatment of the same idea.

By a point-to-point *transformation* $X \to X'$ we mean a rule for associating every point X with every point X' so that there is exactly one X' for each X and exactly one X for each X'. A line-to-line transformation $x \to x'$ is defined similarly. A *collineation* is a point-to-point and line-to-line transformation that preserves the relation of incidence. Thus it transforms ranges into ranges, pencils into pencils, quadrangles into quadrangles, and so on. Clearly, it is a self-dual concept, the inverse of a collineation is a collineation, and the product of two collineations is again a collineation.

A *projective collineation* is a collineation that transforms every one-dimensional form (range or pencil) projectively, so that, if it transforms the points Y on a line b into the points Y' on the corresponding line b', the relation between Y and Y' is a projectivity $Y \barwedge Y'$. The following remarkable theorem is reminiscent of 5.31:

6.11 *Any collineation that transforms one range projectively is a projective collineation.*

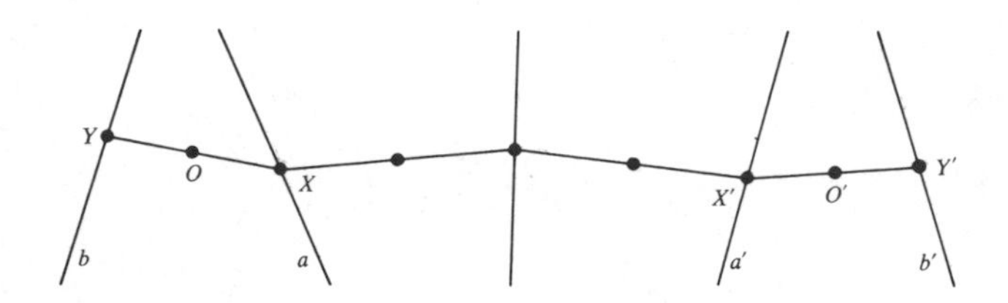

FIGURE 6.1A

PROOF. Let a and a' be the corresponding lines that carry the projectively related ranges. We wish to establish the same kind of relationship between any other pair of corresponding lines, say b and b' (Figure 6.1A). Let Y be a variable point on b, and O a fixed point on neither a nor b. Let OY meet a in X. The given collineation transforms O into a fixed point O' (on neither a' nor b'), and OY into a line $O'Y'$ that meets a' in X'. Since X is on the special line a, we have $X \barwedge X'$. Thus

$$Y \overset{O}{\doublebarwedge} X \barwedge X' \overset{O'}{\doublebarwedge} Y':$$

the collineation induces a projectivity $Y \barwedge Y'$ between b and b', as desired.

To obtain the dual result (with "range" replaced by "pencil") we merely have to regard the ranges Y and Y' as sections of corresponding pencils.

Axiom 2.18 yields the following two-dimensional analogue:

6.12 *The only projective collineation that leaves invariant* 4 *lines forming a quadrilateral, or* 4 *points forming a quadrangle, is the identity.*

PROOF. Suppose the sides of a quadrilateral are 4 invariant lines. Then the vertices (where the sides intersect in pairs) are 6 invariant points, 3 on each side. Since the relation between corresponding sides is projective, every point on each side is invariant. Any other line contains invariant points where it meets the sides and is consequently invariant. Thus the collineation must be the identity. The dual argument gives the same result when there is an invariant quadrangle.

The analogue of the fundamental theorem 4.12 is as follows:

6.13 *Given any two complete quadrilaterals (or quadrangles), with their four sides (or vertices) named in a corresponding order, there is just one projective collineation that will transform the first into the second.*

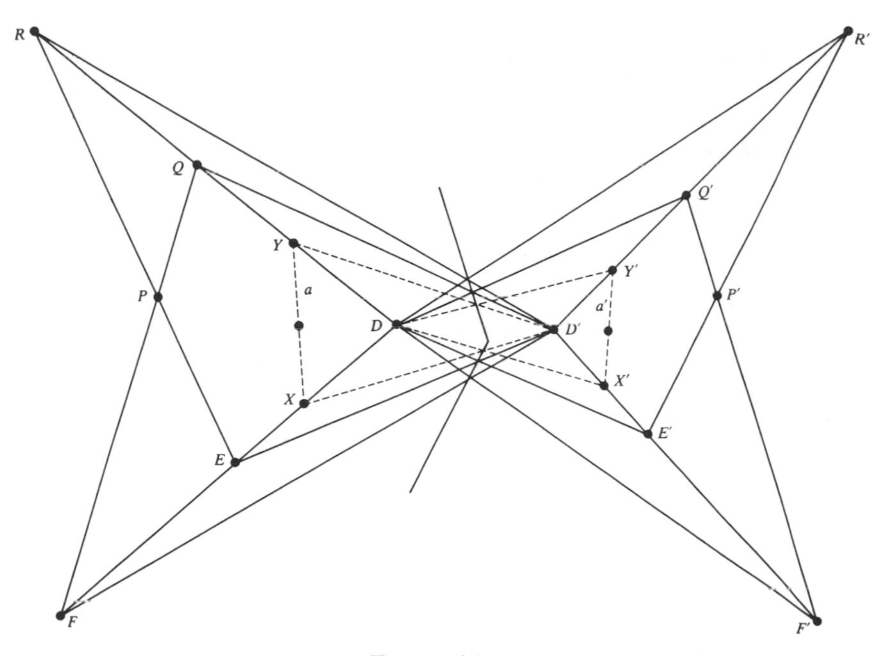

FIGURE 6.1B

PROOF. Let $DEFPQR$ and $D'E'F'P'Q'R'$ be the two given quadrilaterals, as in Figure 6.1B. We proceed to investigate the effect that such a collineation should have on an arbitrary line a. There are certainly two sides of the first quadrilateral that meet a in two distinct points. For definiteness, suppose a is XY, with X on DE and Y on DQ. The projectivities $DEF \barwedge D'E'F'$ and $DQR \barwedge D'Q'R'$ determine a line $a' = X'Y'$, where

$$DEFX \barwedge D'E'F'X' \quad \text{and} \quad DQRY \barwedge D'Q'R'Y'.$$

To prove that this correspondence $a \rightarrow a'$ is a collineation, we have to verify that it also relates points to points in such a way that incidences are preserved. For this purpose, let a vary in a pencil, so that $X \doublebarwedge Y$. By our construction for a', we now have

$$X' \barwedge X \doublebarwedge Y \barwedge Y'.$$

Since D is the invariant point of the perspectivity $X \doublebarwedge Y$, D' must be an invariant point of the projectivity $X' \barwedge Y'$. Hence, by 4.22, this projectivity is again a perspectivity. Thus a', like a, varies in a pencil: that is, concurrent lines yield concurrent lines. We have not only a line-to-line transformation but also a point-to-point transformation, preserving incidences, namely, a collineation. The projectivity $X \barwedge X'$ suffices to make it a *projective* collineation.

Finally, there is no other projective collineation transforming $DEFPQR$ into $D'E'F'P'Q'R'$; for, if another transformed a into a_1, the inverse of the latter would take a_1 to a, the original collineation takes a to a', and altogether we would have a projective collineation leaving $D'E'F'P'Q'R'$ invariant and taking a_1 to a'. By 6.12, this combined collineation can only be the identity. Thus, for every a, a_1 coincides with a': the "other" collineation is really the old one over again. In other words, the projective collineation $a \to a'$ is unique.

In the statement of the theorem, we used the phrase "named in a corresponding order." This was necessary, because otherwise we could have permuted the sides of one of the quadrilaterals in any one of $4! = 24$ ways, obtaining not just one collineation but 24 collineations.

We happened to use quadrilaterals, but the dual argument would immediately yield the same result for quadrangles.

EXERCISE

Let $PQRS \to P'Q'R'S'$ denote the projective collineation that relates two given quadrangles $PQRS$ and $P'Q'R'S'$; for instance, $PQRS \to PQRS$ is the identity. Describe the special collineations

(i) $PQRS \to QPRS$, (ii) $PQRS \to QRPS$, (iii) $PQRS \to QRSP$.

Hints:

(i) What happens to the lines PQ and RS?

(ii) What does Figure 3.4A tell us about the possibility of an invariant line?

(iii) What happens to the lines PR, QS, and to the points $QR \cdot PS$, $PQ \cdot RS$?

6.2 Perspective Collineations

In Section 2.3, we obtained the Desargues configuration, Figure 2.3A, by taking two triangles PQR and $P'Q'R'$, perspective from O. By 6.13, there is just one projective collineation that transforms the quadrangle $DEPQ$ into $DEP'Q'$. This collineation, transforming the line $o = DE$ into itself, and PQ

into $P'Q'$, leaves invariant the point $o \cdot PQ = F = o \cdot P'Q'$. By Axiom 2.18, it leaves invariant every point on o. The join of any two distinct corresponding points meets o in an invariant point, and is therefore an invariant line. The two invariant lines PP' and QQ' meet in an invariant point, namely O.

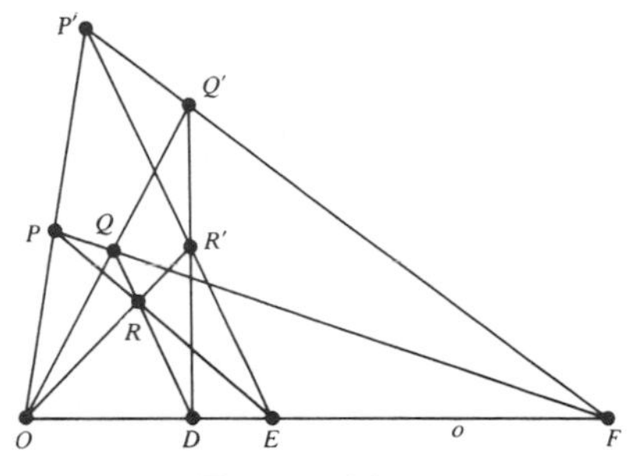

FIGURE 6.2A

The point $R = DQ \cdot EP$ is transformed into $DQ' \cdot EP' = R'$. By the dual of Axiom 2.18, every line through O is invariant.

This collineation, relating two perspective triangles, is naturally called a *perspective collineation*. The point O and line o, from which the triangles are perspective, are the *center* and *axis* of the perspective collineation. If O and o are nonincident, as in Figure 2.3A, the collineation is a *homology* (so named by Poncelet). If O and o are incident, as in Figure 6.2A, it is an *elation* (so named by the Norwegian geometer Sophus Lie, 1842–1899). To sum up,

6.21 *Any two perspective triangles are related by a perspective collineation, namely an elation or a homology according as the center and axis are or are not incident.*

These ideas are further elucidated in the following six theorems.

6.22 *A homology is determined when its center and axis and one pair of corresponding points (collinear with the center) are given.*

PROOF. Let O be the center, o the axis, P and P' (collinear with O) the given corresponding points. We proceed to set up a construction whereby each point R yields a definite corresponding point R'. If R coincides with O or lies on o, it is, of course, invariant, that is, R' coincides with R. If, as in Figure 2.3A, R is neither on o nor on OP, we have

$$R' = EP' \cdot OR, \quad \text{where} \quad E = o \cdot PR.$$

Finally, if R is on OP, as in Figure 6.2B, we use an auxiliary pair of points Q, Q' (of which the former is arbitrary while the latter is derived from it the way we just now derived R' from R) to obtain

$$R' = DQ' \cdot OP, \quad \text{where} \quad D = o \cdot QR.$$

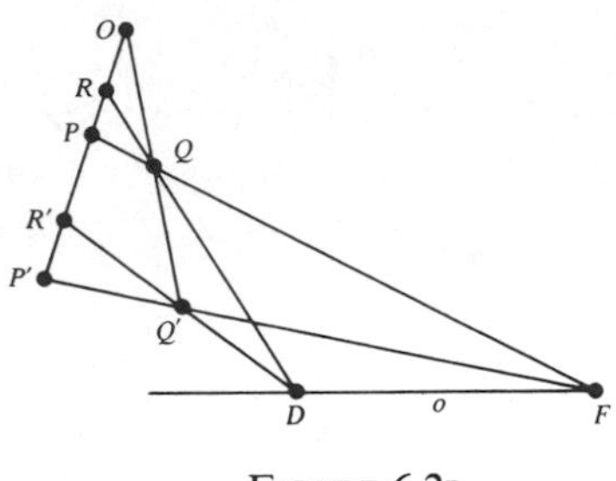

FIGURE 6.2B

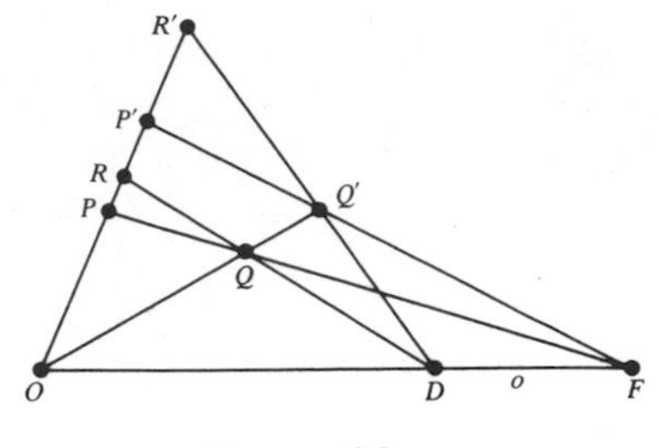

FIGURE 6.2C

6.23 *An elation is determined when its axis and one pair of corresponding points are given.*

PROOF. Let o be the axis, P and P' the given pair. Since the collineation is known to be an elation, its center is $o \cdot PP'$. We proceed as in the proof of 6.22, using Figures 6.2A and 6.2C instead of 2.3A and 6.2B.

This elation, with center $o \cdot PP'$, is conveniently denoted by $[o; P \to P']$.

6.24 *Any collineation that has one range of invariant points (but not more than one) is perspective.*

PROOF. Since the identity is (trivially) a projectivity, 6.11 tells us that any such collineation is projective. There cannot be more than one invariant point outside the line o whose points are all invariant; for, two such would form, with two arbitrary points on o, a quadrangle of the kind considered in 6.12. If there is one invariant point O outside o, every line through O meets o in another invariant point; that is, every line through O is invariant. Any noninvariant point P lies on such a line and is therefore transformed into another point P' on this line OP; hence the collineation is a homology, as in 6.22. If, on the other hand, all the invariant points lie on o, any two distinct joins PP' and QQ' (of pairs of corresponding points) must meet o in the same point O, and the collineation is an elation, as in 6.23.

Hence, also,

6.25 *If a collineation has a range of invariant points, it has a pencil of invariant lines.*

6.26 *All the invariant points of an elation lie on its axis.*

6.27 *For a homology, the center is the only invariant point not on the axis.*

EXERCISES

1. Let P, P', Q, Q', D be five points, no three collinear. Then there is a unique perspective collineation that takes P to P', and Q to Q', while its axis passes through D.

2. What kind of projectivity does a perspective collineation induce on a line through its center?
3. If, on a line through its center O, an elation transforms A into A', and B into B', then $(OO)(AB')(A'B)$. (Reference **19**, p. 78.) [*Hint:* Use 5.11.]
4. What kind of transformation is the product of two elations having the same axis?
5. Elations with a common axis are commutative.
6. What kind of projective collineation will leave invariant two points and two lines (the points not lying on the lines)?
7. Referring to Theorem 6.12, discuss the significance of the phrase "forming a quadrilateral."
8. When the projective plane is regarded as an extension of the Euclidean plane, what is the Euclidean name for a perspective collineation whose axis is the line at infinity?

6.3 Involutory Collineations

Suppose a given transformation relates a point X to X', X' to X'', X'' to X''', . . . , $X^{(n-1)}$ to $X^{(n)}$. If, for every position of X, $X^{(n)}$ coincides with X itself, the transformation is said to be *periodic* and the smallest n for which this happens is called the *period.* Thus the identity is of period 1, an involution is (by definition) of period 2, and the projectivity $ABC \barwedge BCA$ (for any three distinct collinear points A, B, C) is of period 3.

We saw, in 6.22, that a homology is determined by its center O, axis o, and one pair of corresponding points P, P'. In the special case when the harmonic conjugate of O with respect to P and P' lies on o, we speak of a *harmonic* homology. Clearly

6.31 *A harmonic homology is determined when its center and axis are given.*

For any point P, the corresponding point P' is simply the harmonic conjugate of P with respect to O and $o \cdot OP$. Thus a harmonic homology is of period 2. Conversely,

6.32 *Every projective collineation of period* 2 *is a harmonic homology.*

PROOF. Given a projective collineation of period 2, suppose it interchanges the pair of distinct points PP' and also another pair QQ' (not on the line PP'). By 6.13, it is the only projective collineation that transforms the

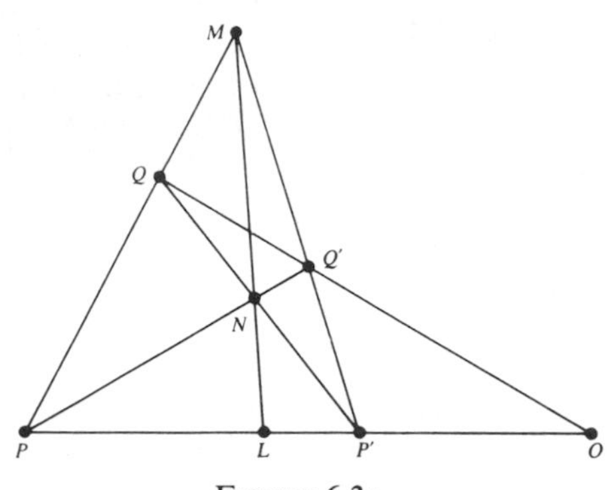

FIGURE 6.3A

quadrangle $PP'QQ'$ into $P'PQ'Q$. The invariant lines PP' and QQ' meet in an invariant point O, as in Figure 6.3A. Since the collineation interchanges the pair of lines PQ, $P'Q'$, and likewise the pair PQ', $P'Q$, the two points

$$M = PQ \cdot P'Q' \quad \text{and} \quad N = PQ' \cdot P'Q$$

are invariant. Moreover, the two invariant lines PP' and MN meet in a third invariant point L on MN. By Axiom 2.18, every point on MN is invariant. Thus the collineation is perspective (according to the definition in Section 6.2). Since, by Axiom 2.17, its center O does not lie on its axis MN, it is a homology. Finally, since $\mathsf{H}(PP', OL)$, it is a harmonic homology.

EXERCISES

1. What kind of transformation is the product of two harmonic homologies having the same axis but different centers?
2. Any elation with axis o may be expressed as the product of two harmonic homologies having this same axis o.
3. What kind of collineation is the product of three harmonic homologies whose centers and axes are the vertices and opposite sides of a triangle?
4. The product of two harmonic homologies is a homology if and only if the center of each lies on the axis of the other. In this case the product is again a harmonic homology. (Reference **7**, pp. 64–65.)
5. What is the Euclidean name for a harmonic homology whose axis is the line at infinity?

6.4 Projective Correlations

We have already considered a simple instance of a point-to-line transformation: the elementary correspondence that relates a range to a pencil when the former is a section of the latter. We shall now extend this to a transformation $X \to x'$ relating all the points in a plane to all the lines in the same plane, and its dual $x \to X'$ which relates all the lines to all the points. A *correlation* is a point-to-line and line-to-point transformation that preserves the relation of incidence in accordance with the principle of duality. Thus it transforms ranges into pencils, pencils into ranges, quadrangles into quadrilaterals, and so on. A correlation is a self-dual concept, the inverse of a correlation is again a correlation, and the product of two correlations is a collineation.

A *projective* correlation is a correlation that transforms every one-dimensional form projectively, so that, if it transforms the points Y on a line b into the lines y' through the corresponding point B', the relation between Y and y' is a projectivity $Y \barwedge y'$. There is a theorem analogous to 6.11:

6.41 *Any correlation that transforms one range projectively is a projective correlation.*

PROOF. Let a and A' be the corresponding line and point that carry the projectively related range and pencil $X \barwedge x'$. We wish to establish the same kind of relationship between any other corresponding pair, say b and B' (see Figure 6.4A). Let Y be a variable point on b, and O a fixed point on neither a nor b. Let OY meet a in X. The given correlation transforms O into a fixed line o' (through neither A' nor B'), and OY into a point $o' \cdot y'$ which is joined to A' by a line x'. Since

$$Y \overset{O}{\barwedge} X \barwedge x' \overset{o'}{\barwedge} y',$$

the correlation induces a projectivity $Y \barwedge y'$ between b and B' as desired. To obtain the dual result for a pencil and the corresponding range, we

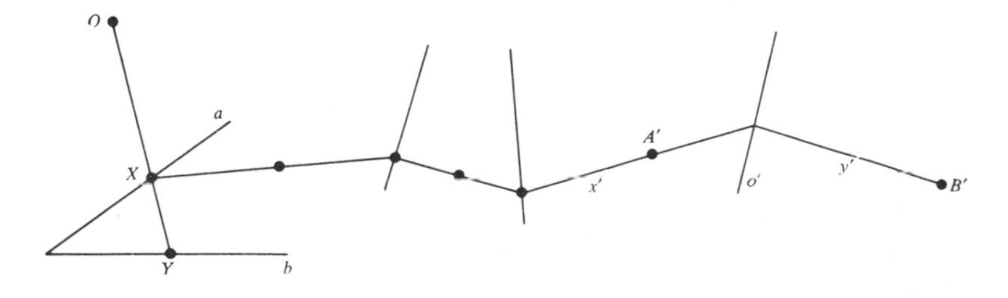

FIGURE 6.4A

merely have to regard the range of points Y on b as a section of the given pencil. This pencil yields a range which is a section of the pencil of lines y' through B'.

As a counterpart for 6.13, we have:

6.42 *A quadrangle and a quadrilateral, with the four vertices of the former associated in a definite order with the four sides of the latter, are related by just one projective correlation.*

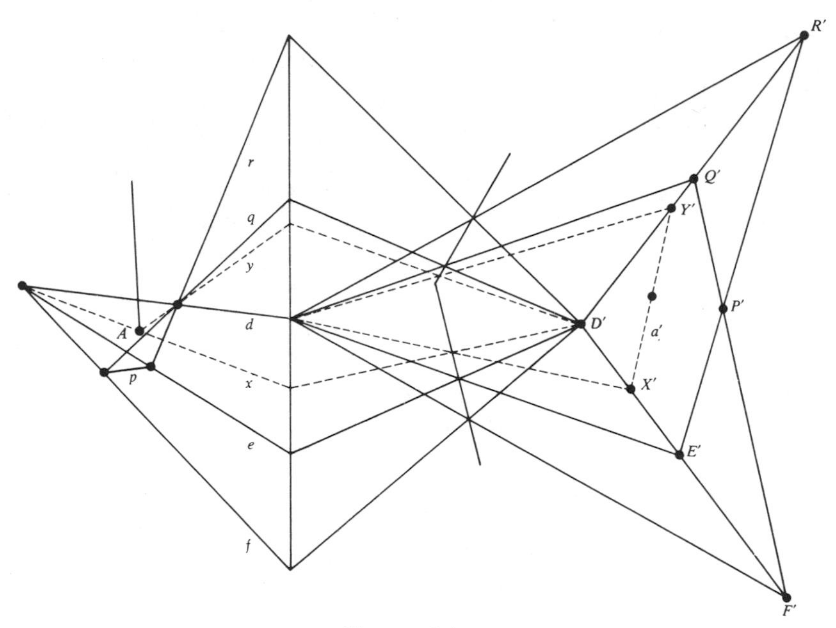

FIGURE 6.4B

PROOF. Let $defpqr$ and $D'E'F'P'Q'R'$ be the given quadrangle and quadrilateral, as in Figure 6.4B. What effect should such a correlation have on an arbitrary point A? For definiteness suppose A is $x \cdot y$ with x through $d \cdot e$ and y through $d \cdot q$. The projectivities $def \barwedge D'E'F'$ and $dqr \barwedge D'Q'R'$ determine a line $a' = X'Y'$, where

$$defx \barwedge D'E'F'X', \quad dqry \barwedge D'Q'R'Y'.$$

To prove that this correspondence $A \rightarrow a'$ is a correlation, we have to verify that it also relates lines to points in such a way that incidences are preserved. For this purpose, let A vary in a range, so that $x \doublebarwedge y$. By our construction for a', we now have

$$X' \barwedge x \doublebarwedge y \barwedge Y'.$$

Since d is an invariant line of the perspectivity $x \;\overline{\overline{\wedge}}\; y$, D' must be an invariant point of the projectivity $X' \;\overline{\wedge}\; Y'$. Thus a' varies in a pencil; that is, collinear points yield concurrent lines. We have not only a point-to-line transformation but also a line-to-point transformation, dualizing incidences, namely, a correlation. The projectivity $x \;\overline{\wedge}\; X'$ suffices to make it a projective correlation.

Finally, there is no other projective correlation transforming $defpqr$ into $D'E'F'P'Q'R'$; for, if another transformed A into a_1, the inverse of the latter would take a_1 to A, the original correlation takes A to a', and altogether we would have a projective collineation leaving $D'E'F'P'Q'R'$ invariant and taking a_1 to a'. As in Section 6.1, this establishes the uniqueness of the correlation $A \rightarrow a'$.

The dual construction yields a projective correlation transforming a given quadrilateral into a given quadrangle.

EXERCISE

If a correlation transforms a given quadrangle into a quadrilateral, it transforms the three diagonal points of the quadrangle into the three diagonal lines of the quadrilateral.

CHAPTER SEVEN

Polarities

Chasles was the last important member of the great French school of projective geometers. After his time primacy in this subject passed across the Rhine, never to return.

J. L. Coolidge (Reference **4**, p. 58)

7.1 Conjugate Points and Conjugate Lines

A *polarity* is a projective correlation of period 2. In general, a correlation transforms each point A into a line a' and transforms this line into a new point A''. When the correlation is of period 2, A'' always coincides with A and we can simplify the notation by omitting the prime ($'$). Thus a polarity relates A to a, and vice versa. Following J. D. Gergonne (1771–1859), we call a the *polar* of A, and A the *pole* of a. Since this is a projective correlation, the polars of all the points on a form a projectively related pencil of lines through A.

Since a polarity dualizes incidences, if A lies on b, the polar a passes through the pole B. In this case we say that A and B are *conjugate points*, and that a and b are *conjugate lines*. It may happen that A and a are incident, so that each is *self-conjugate*: A on its own polar, and a through its own pole. However, the occurrence of self-conjugate lines (and points) is restricted by the following theorem:

7.11 *The join of two self-conjugate points cannot be a self-conjugate line.*

PROOF. If the join a of two self-conjugate points were a self-conjugate line, it would contain its own pole A and at least one other self-conjugate point, say B. The polar of B, containing both A and B, would coincide with

a: two distinct points would both have the same polar. This is impossible, since a polarity is a one-to-one correspondence between points and lines.

The occurrence of self-conjugacy is further restricted as follows:

7.12 *It is impossible for a line to contain more than two self-conjugate points.*

PROOF. Let p and q (through C) be the polars of two self-conjugate points P and Q on a line c, as in Figure 7.1A. Let R be a point on p, distinct from C and P. Let its polar r meet q in S. Then $S = q \cdot r$ is the pole of $QR = s$, which meets r in T, say. Also $T = r \cdot s$ is the pole of $RS = t$,

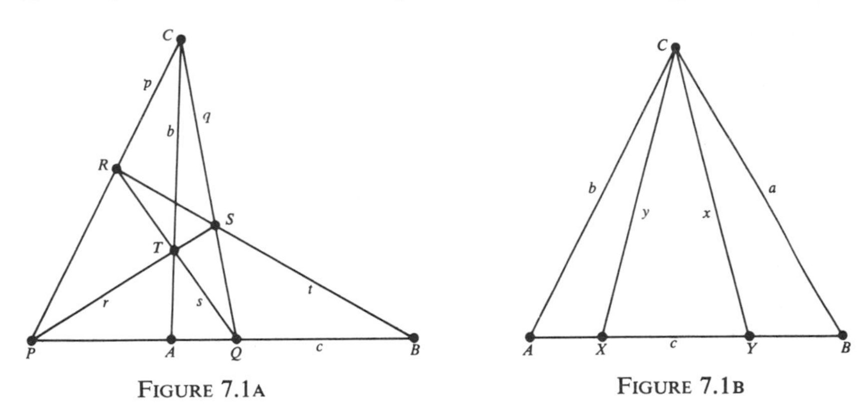

FIGURE 7.1A FIGURE 7.1B

which meets c in B, say. Finally, $B = c \cdot t$ is the pole of $CT = b$, which meets c in A, the harmonic conjugate of B with respect to P and Q.

The point B cannot coincide with Q or P. For, $B = Q$ would imply $R = C$; and $B = P$ would imply $S = C, r = p, R = P$; but we are assuming that R is neither C nor P. Hence, by 2.51, $A \neq B$, and B is not self-conjugate. We thus have, on c, two self-conjugate points P, Q and a non-selfconjugate point B.

Since the polars of a range form a projectively related pencil, each point X on c determines a conjugate point Y on c, which is where its polar x meets c (see Figure 7.1B), and this correspondence between X and Y is a projectivity:

$$X \barwedge x \barwedge Y.$$

When X is P, x is p, and Y is P again; thus P is an invariant point of this projectivity. Similarly, Q is another invariant point. But when X is B, Y is the distinct point A; therefore the projectivity is not the identity. By Axiom 2.18, P and Q are its *only* invariant points; that is, P and Q are the only self-conjugate points on c. This completes the proof that c cannot contain more than two self-conjugate points.

A closely related result is this:

7.13 *A polarity induces an involution of conjugate points on any line that is not self-conjugate.*

PROOF. On a non-selfconjugate line c, the projectivity $X \barwedge Y$, where $Y = c \cdot x$ (as in Figure 7.1B), transforms any non-selfconjugate point B into another point $A = b \cdot c$, whose polar is BC. The same projectivity transforms A into B. Since it interchanges A and B, it must be an involution.

Dually, the lines x and CX are paired in the involution of conjugate lines through C.

Such a triangle ABC, in which each vertex is the pole of the opposite side (so that any two vertices are conjugate points, and any two sides are conjugate lines), is called a *self-polar triangle.*

EXERCISES

1. Every self-conjugate line contains just one self-conjugate point.
2. If a and b are nonconjugate lines, every point X on a has a conjugate point Y on b. The relation between X and Y is a projectivity. It is a perspectivity if and only if the point $a \cdot b$ is self-conjugate. (Reference **19,** p. 124.) [*Hint:* Use 4.22.]
3. Observe that the polarity of 7.12 (Figure 7.1A) relates the quadrangle $CRST$ to the quadrilateral $crst$, and the harmonic set of points P, Q, A, B to the harmonic set of lines CP, CQ, CB, CA.
4. In the polarity of 7.12, find the poles of RA and SA.
5. In Section 3.4 we defined a "trilinear polarity." Is this a true polarity?

7.2 The Use of a Self-Polar Triangle

We have referred to 6.11 as an analogue of 5.31. Another possible candidate for the same distinction is as follows:

7.21 *Any projective correlation that relates the three vertices of one triangle to the respectively opposite sides is a polarity.*

PROOF. Consider the correlation $ABCP \rightarrow abcp$, where a, b, c are the sides of the given triangle ABC and P is a point not on any of them. Then p is a line not through any of A, B, C. The point P and line p determine 6 points on the sides of the triangle, as in Figure 7.2A:

$$P_a = a \cdot AP,\ P_b = b \cdot BP,\ P_c = c \cdot CP,\ A_p = a \cdot p,\ B_p = b \cdot p,\ C_p = c \cdot p.$$

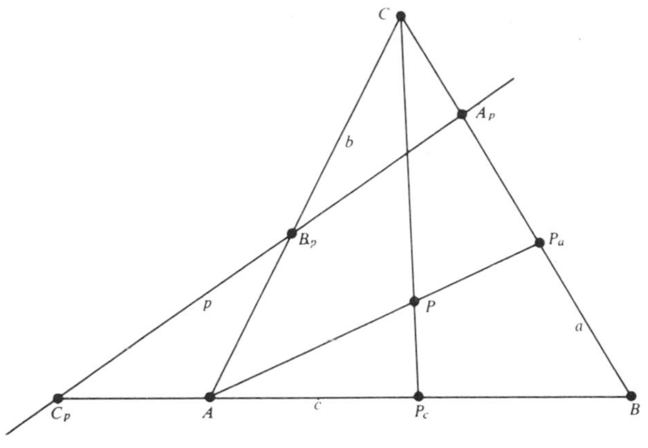

FIGURE 7.2A

The correlation, transforming A, B, C into a, b, c, also transforms $a = BC$ into $b \cdot c = A$, AP into $a \cdot p = A_p$, $P_a = a \cdot AP$ into AA_p, and so on. Thus it transforms the triangle ABC in the manner of a polarity, but we still have to investigate whether, besides transforming P into p, it also transforms p into P.

The correlation transforms each point X on c into a certain line which intersects c in Y, say. Since it is a projective correlation, we have $X \barwedge Y$. When X is A, Y is B; and when X is B, Y is A. Thus the projectivity $X \barwedge Y$ interchanges A and B, and is an involution. Since the correlation transforms P_c into CC_p, the involution includes P_cC_p as one of its pairs. Hence the correlation transforms C_p into CP_c, which is CP. Similarly, it transforms A_p into AP, and B_p into BP. Therefore it transforms $p = A_pB_p$ into $AP \cdot BP = P$, as required.

We have now proved that the correlation $ABCP \rightarrow abcp$ is a polarity. An appropriate symbol, analogous to the symbol $(AB)(PQ)$ for an involution, is

$$(ABC)(Pp).$$

Thus any triangle ABC, any point P not on a side, and any line p not through a vertex, determine a definite polarity $(ABC)(Pp)$, in which the polar x of an arbitrary point X can be constructed by incidences.

This construction could be carried out by the method of 6.42, as applied to the correlation $ABCP \rightarrow abcp$. More elegantly, we could adapt the notation of Figure 7.2A so that

$$X_a = a \cdot AX, \quad X_b = b \cdot BX, \quad A_x = a \cdot x, \quad B_x = b \cdot x.$$

Then A_x is the mate of X_a in the involution $(BC)(P_aA_p)$, B_x is the mate of X_b in $(CA)(P_bB_p)$, and x is A_xB_x. A still simpler procedure will be described in Section 7.4, but it seems desirable to deal with some other matters first.

Consider a polarity $(ABC)(Pp)$, in which P does not lie on p (see Figure 7.2A). Since the polars of the points

$$A_p = a \cdot p, \quad B_p = b \cdot p, \quad C_p = c \cdot p$$

are AP, BP, CP, the pairs of opposite sides of the quadrangle $ABCP$ meet the line p in pairs of *conjugate* points. Hence

7.22 *In a polarity $(ABC)(Pp)$, where P is not on p, the involution of conjugate points on p is the involution determined on p by the quadrangle $ABCP$.*

EXERCISES

1. In the notation of Figure 2.4A, any projective correlation that relates the points S, D, E, F to the respective lines g, SA, SB, SC is a polarity.
2. Consider a polarity $(ABC)(Pp)$ in which P *is* on p. Find Q and q, *not* incident, so that the same polarity can be described as $(ABC)(Qq)$.

7.3 Polar Triangles

From any given triangle we can derive a *polar triangle* by taking the polars of the three vertices, or the poles of the three sides. M. Chasles observed that a triangle and its polar triangle, if distinct, are perspective. In other words,

7.31 CHASLES'S THEOREM: *If the polars of the vertices of a triangle do not coincide with the respectively opposite sides, they meet these sides in three collinear points.*

PROOF. Let PQR be a triangle whose sides QR, RP, PQ meet the polars p, q, r of its vertices in points P_1, Q_1, R_1, as in Figure 7.3A. The polar of $R_1 = PQ \cdot r$ is, of course, $r_1 = (p \cdot q)R$. Define also the extra points $P' = PQ \cdot q$, $R' = QR \cdot q$, and the polar $p' = (p \cdot q)Q$ of the former. By Theorem 1.63 and the polarity, we have

$$R_1PP'Q \barwedge PR_1QP' \barwedge pr_1qp' \barwedge P_1RR'Q.$$

By 4.22 (since Q is invariant), $R_1PP' \overline{\barwedge} P_1RR'$. The center of the perspectivity, namely $PR \cdot P'R' = Q_1$, must lie on the line R_1P_1. Hence P_1, Q_1, R_1 are collinear, as desired.

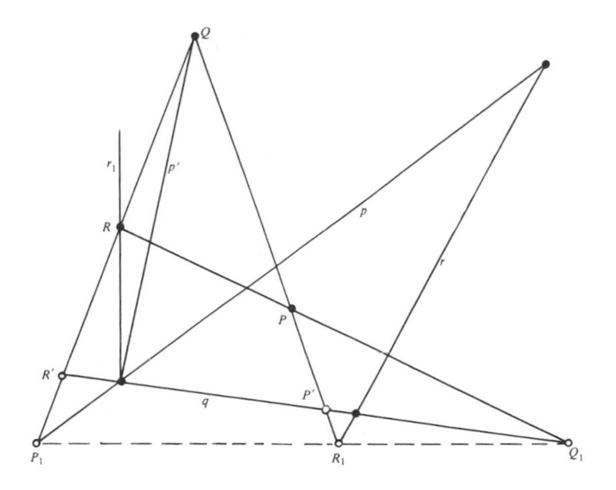

FIGURE 7.3A

This proof breaks down if P_1 or Q lies on q. In the former case, P_1 $(=R')$ and R_1 $(=P')$ are collinear with Q_1. In the latter (when Q lies on q) we can permute the names of P, Q, R (and correspondingly p, q, r), or call the first triangle pqr and the second PQR, in such a way that the new Q and q are *not* incident. It is evidently impossible for *each* triangle to be inscribed in the other.

EXERCISES

1. A triangle and its polar triangle (if distinct) are perspective from a line, and therefore also from a point. The point is the pole of the line.
2. Two triangles ABP and abp, with A on b, B on a, but no other incidences, are polar triangles for a unique polarity.

7.4 A Construction for the Polar of a Point

We are now ready to describe the "still simpler procedure" (Figure 7.4A) which was promised in Section 7.2.

7·41 *The polar of a point X (not on AP, BP, or p) in the polarity $(ABC)(Pp)$ is the line X_1X_2 determined by*

$$A_1 = a \cdot PX, \quad P_1 = p \cdot AX, \quad X_1 = AP \cdot A_1P_1,$$
$$B_2 = b \cdot PX, \quad P_2 = p \cdot BX, \quad X_2 = BP \cdot B_2P_2.$$

PROOF. Applying 7.31 to the triangle PAX, we deduce that its sides AX, XP, PA meet the polars p, a, x of its vertices in three collinear points,

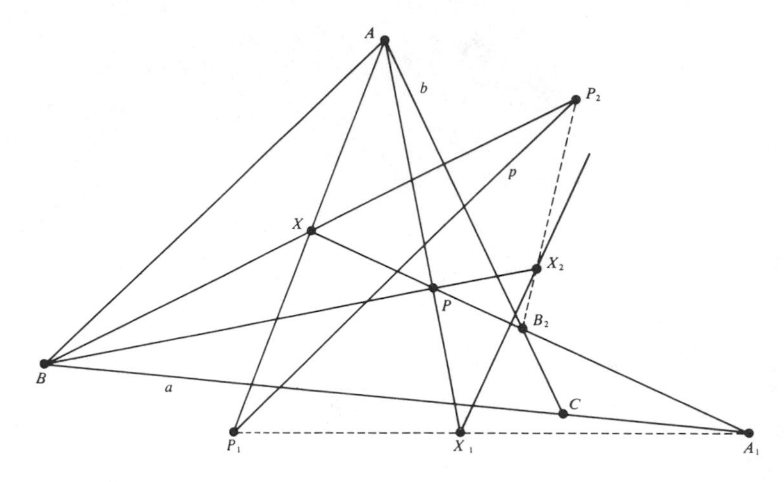

FIGURE 7.4A

the first two of which are P_1 and A_1. Hence x must meet PA in a point lying on P_1A_1, namely, in the point $PA \cdot P_1A_1 = X_1$. That is, x passes through X_1. Similarly (by applying 7.31 to the triangle PBX instead of PAX), x passes through X_2.

This construction fails when X lies on AP, for then A_1P_1 coincides with AP, and X_1 is no longer properly defined. However, since X_2 can still be constructed as above, the polar of X is now A_pX_2 (where $A_p = a \cdot p$). Similarly, when X is on BP, its polar is X_1B_p.

Finally, to locate the polar of a point X on p, we can apply the dual of the above construction to locate the pole Y of a line y through X. This y may be any line through X except p or PX. (It is convenient to choose $y = AX$ or, if this happens to coincide with PX, to choose $y = BX$.) Then the desired polar is $x = PY$.

EXERCISES

1. For any point X, not on AP, BP, or p, the polar is

$$[AP \cdot (a \cdot PX)(p \cdot AX)][BP \cdot (b \cdot PX)(p \cdot BX)].$$

Write out the dual expression for the pole of any line x, not through A_p, B_p, or P, and draw a figure to illustrate this dual construction.

2. If X lies on AB, x joins C to

$$AP \cdot (a \cdot PX)(p \cdot c).$$

Deduce an alternative construction for the pole of a line WX, with W on CA and X on AB.

7.5 The Use of a Self-Polar Pentagon

Instead of describing a polarity as $(ABC)(Pp)$, we can equally well describe it in terms of a *self-polar pentagon*, that is, a pentagon in which each of the five vertices is the pole of the "opposite" side, as in Figure 7.5A. This way of specifying a polarity is a consequence of the following theorem, due to von Staudt:

7.51 *The projective correlation that transforms four vertices of a pentagon into the respectively opposite sides is a polarity and transforms the remaining vertex into the remaining side.*

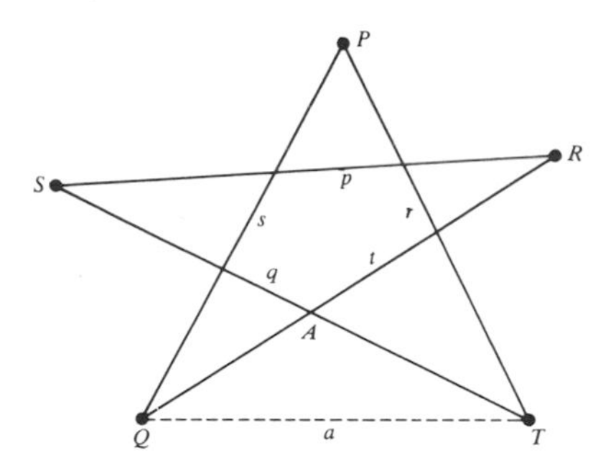

FIGURE 7.5A

PROOF. The correlation that transforms the four vertices Q, R, S, T of the pentagon $PQRST$ into the four sides $q = ST$, $r = TP$, $s = PQ$, $t = QR$ also transforms the three sides $t = QR$, $p = RS$, $q = ST$ into the three vertices $T = q \cdot r$, $P = r \cdot s$, $Q = s \cdot t$, and the "diagonal point" $A = q \cdot t$ into the "diagonal line" $a = QT$. Thus it transforms each vertex of the triangle AQT into the opposite side. By 7.21, it is a polarity, namely (since it transforms p into P), the polarity $(AQT)(Pp)$.

EXERCISES

1. In the notation of Figure 7.2A, PBA_pB_pA is a self-polar pentagon.

2. Let X be any point on none of the sides p, r, s of a given self-polar pentagon $PQRST$. Then its polar is the line

$$[r \cdot (t \cdot PX)(p \cdot TX)][s \cdot (q \cdot PX)(p \cdot QX)].$$

7.6 A Self-Conjugate Quadrilateral

From Chasles's theorem, 7.31, we can easily deduce

7.61 HESSE'S THEOREM: *If two pairs of opposite vertices of a complete quadrilateral are pairs of conjugate points* (*in a given polarity*), *then the third pair of opposite vertices is likewise a pair of conjugate points.*

PROOF. Consider a quadrilateral $PQRP_1Q_1R_1$, as in Figure 7.3A, with P conjugate to P_1, and Q to Q_1. The polars p and q (of P and Q) pass through P_1 and Q_1, respectively. By Chasles's theorem (7.31) the polar of R meets PQ in a point that lies on P_1Q_1, namely in the point $PQ \cdot P_1Q_1 = R_1$. Therefore the polar of R passes through R_1; that is, R is conjugate to R_1.

EXERCISE

In the notation of Figure 7.5A, let an arbitrary line through P meet ST in U, and QR in V. Then RU and SV are conjugate lines.

7.7 The Product of Two Polarities

Figure 2.3A (or Figure 6.2B) shows the homology with center O and axis $o = DF$ that transforms P into P' (and consequently Q into Q'). This homology may be expressed as the product of two polarities

$$(ODF)(Pp) \quad \text{and} \quad (ODF)(P'p),$$

where p is any line not passing through a vertex of the common self-polar triangle ODF. To prove this, we merely have to observe that the homology and the product of polarities both transform the quadrangle $ODFP$ into $ODFP'$.

Unfortunately, this expression for a homology as the product of two polarities cannot in any simple way be adapted to an elation. Accordingly, it is worthwhile to mention a subtler expression that applies equally well to either kind of perspective collineation. Figure 7.7A shows the homology or elation with center O and axis $o = CP$ that transforms A into another point A' on the line $c = OA$. Here C and P are arbitrary points on the axis o (which passes through O if the collineation is an elation). Let p be an arbitrary line through O, meeting $b = CA$ and $b' = CA'$ in Q and Q'. Let B be an arbitrary point on c. We proceed to verify that the given perspective collineation is the product of the two polarities

$$(ABC)(Pp) \quad \text{and} \quad (A'BC)(Pp).$$

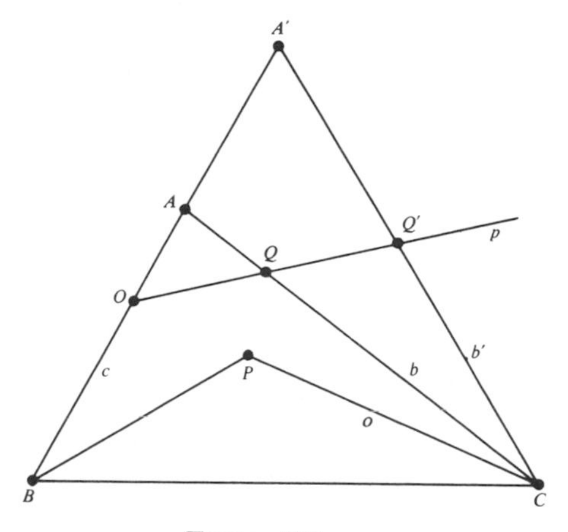

FIGURE 7.7A

In fact, the first polarity transforms the four points A, P, $O = c \cdot p$, $Q = b \cdot p$ into the four lines BC, p, CP, BP; and the second transforms these lines into the four points A', P, $c \cdot p = O$, $b' \cdot p = Q'$. Thus their product transforms the quadrangle $APOQ$ into $A'POQ'$. By 6.13, this product is the same as the given perspective collineation.

More generally,

7.71 *Any projective collineation is expressible as the product of two polarities.*

PROOF. By the above remarks, this is certainly true if the given collineation is perspective. Accordingly, we may concentrate our attention on a given nonperspective collineation. Let A be a noninvariant point, and l a noninvariant line through A. Suppose the given collineation transforms A into A', A' into A'', l into l', l' into l'', and l'' into l'''. Since the collineation is not perspective, we may choose A and l (as in Figure 7.7B) so that AA' is not an invariant line and $l \cdot l'$ is not an invariant point and so that A'' does not lie on l, nor A' on any of the three lines l, l'', l''', and consequently A

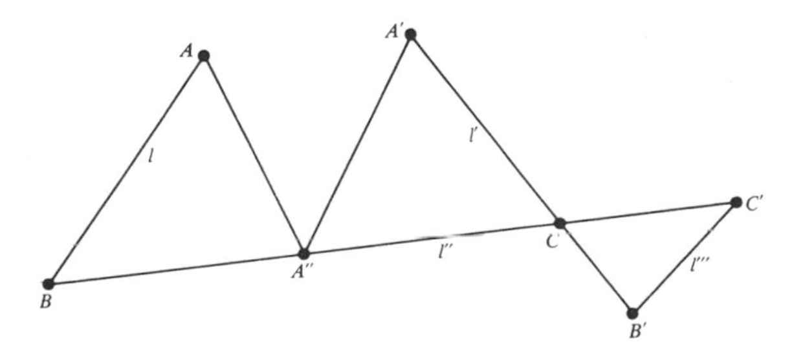

FIGURE 7.7B

does not lie on l' nor on l''. Let l'' meet l in B, l' in C. Of the two polarities

$$(AA''B)(A'l') \quad \text{and} \quad (A'A''C)(Al'''),$$

the former transforms the four points A, A', B, $C = l' \cdot l''$ into the four lines $A''B = l'' = A''C$, $l' = CA'$, $A''A$, $A'A$, and the latter transforms these lines into the four points A', A'', $l' \cdot l''' = B'$, $l'' \cdot l''' = C'$. Hence their product is the same as the given collineation, which is what we wished to prove.

This theorem has an interesting corollary which includes 6.25 as a special case:

7.72 *In any projective collineation, the invariant points and invariant lines form a self-dual figure.*

EXERCISES

1. Can a projective collineation interchange two points without being a harmonic homology?
2. If a projective collineation has three invariant points forming a triangle, it is the product of two polarities having a common self-polar triangle.

7.8 The self-polarity of the Desargues configuration

The Desargues configuration 10_3 can be regarded as a pair of mutually inscribed pentagons, such as *FDROP'* and *EPQQ'R'* (see Figure 2.3A and Section 3.4, Exercise 2). Any pentagon determines a polarity (Section 7.5) for which each vertex is the pole of the opposite side. Consider the polarity for which *FDROP'* is such a self-polar pentagon, having sides

$$f = RO, \quad d = OP', \quad r = P'F, \quad o = FD, \quad p' = DR.$$

Since d passes through A, and f through C, the involution of pairs of conjugate points on o is $(AD)(CF)$. The quadrangle *OPQR* yields the quadrangular relation $(AD)(BE)(CF)$ and thus indicates that e (the polar of E) is OB. Since Q' is $r \cdot e$, q' is RE; since P is $d \cdot q'$, p is DQ'; since R' is $f \cdot p$, r' is FP; and since Q is $p' \cdot r'$, q is $P'R'$. Thus *EPQQ'R'* is another self-polar pentagon. Also the perspective triangles *PQR* and *P'Q'R'* are polar triangles (as in Section 7.3). (This converse of Chasles's theorem was discovered by von Staudt. See Reference **7**, p. 75, Exercise 4.) In the notation of Section 3.4 and the frontispiece,

There is a unique polarity for which G_{ij} is the pole of g_{ij}.

CHAPTER EIGHT

The Conic

> Let us now pause to note that we have swung through a complete circle, from Desargues and Poncelet who started with a conic and defined a polar system, to von Staudt, who starts with a polar system and reaches a conic.
>
> *J. L. Coolidge* (Reference **4**, p. 64)

8.1 How a Hyperbolic Polarity Determines a Conic

The study of conics is said to have begun in 430 B.C., when the Athenians, suffering from a plague, appealed to the oracle at Delos and were told to double the size of Apollo's cubical altar. Attempts to follow this instruction by placing an equal cube beside the original one, or by doubling the edge length (and thus producing a cube of eight times the volume of the original one), made the pestilence worse than ever. At last, Hippocrates of Chios explained that what was needed was to multiply the edge length by the cube root of 2. The first geometrical solution for this problem was given by Archytas about 400 B.C. (Reference **2**, p. 329). He used a twisted curve. Menaechmus, about 340 B.C., found a far simpler solution by the use of conics. For the next six or seven centuries, conics were investigated in great detail, especially by Apollonius of Perga (262–200 B.C.), who coined the names *ellipse*, *parabola*, and *hyperbola*. Interest in this subject was revived in the seventeenth century (A.D.) when Kepler showed how a parabola is at once a limiting case of an ellipse and of a hyperbola, and Blaise Pascal (1623–1662) discovered a projective property of a circle which consequently holds just as well for any kind of conic. We shall see how this characteristic property of a conic can be deduced from an extraordinarily natural and symmetrical definition which was given by von Staudt in his *Geometrie der Lage* (1847). We shall find that, in the projective plane, there is only one kind of conic: the familiar distinction between the ellipse, parabola, and hyperbola can only be made by assigning a special role to the line at infinity.

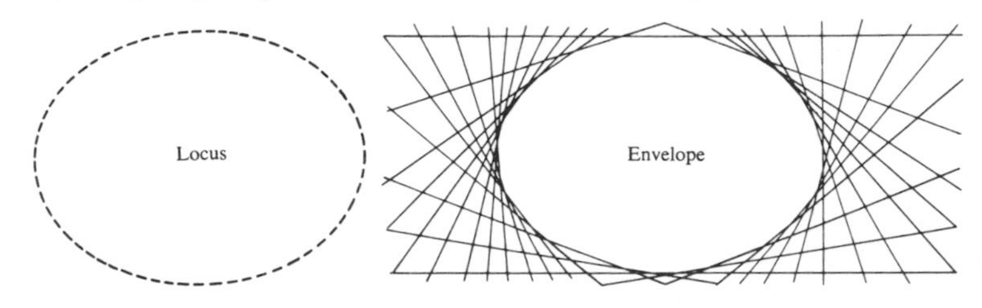

FIGURE 8.1A

Polarities, like involutions, are of two possible kinds. By analogy with involutions, we call a polarity *hyperbolic* or *elliptic* according as it does or does not admit a self-conjugate point. In the former case it also admits a self-conjugate line: the polar of the point. Thus any hyperbolic polarity can be described by a symbol $(ABC)(Pp)$, where P lies on p. This self-conjugate point P, whose existence suffices to make the polarity hyperbolic, is by no means the *only* self-conjugate point: *there is another on every line through* P *except its polar* p.

This can be proved as follows. By 7.11, the only self-conjugate point on a self-conjugate line is its pole. Dually, the only self-conjugate line through a self-conjugate point P is its polar p. By 7.13, it follows that every line through P, except p, is the kind of line that contains an involution of conjugate points. By 5.41, this involution, having one invariant point P, has a second invariant point Q which is, of course, another self-conjugate point of the polarity.

Thus the presence of one self-conjugate point implies the presence of many. Their locus is a *conic*, and their polars are its *tangents*. This simple definition exhibits the conic as a self-dual figure: the locus of self-conjugate points and also the envelope of self-conjugate lines (Figure 8.1A).

In some geometries (such as complex geometry) every polarity is hyperbolic, that is, every polarity determines a conic. In other geometries (e.g., in real geometry) both kinds of polarity occur, and then the theory of polarities is more general than the theory of conics. From here on, we shall deal solely with hyperbolic polarities, so that "pole" will mean "pole with respect to a conic." Similarly, instead of "conjugate for a polarity" we shall say "conjugate with respect to a conic."

A tangent justifies its name by meeting the conic only at its pole: the *point of contact*. Any other line is called a *secant* or a *nonsecant* according as it meets the conic twice or not at all, that is, according as the involution of conjugate points on it is hyperbolic or elliptic. The above remarks show that, of the lines through any point P on the conic, one (namely p) is a tangent and all the others are secants. Moreover, if P and Q are any two distinct points on the conic, the line PQ is a secant.

Dually, a point not lying on the conic is said to be *exterior* or *interior* according as it lies on two tangents or on none, that is, according as the involution of conjugate lines through it is hyperbolic or elliptic. Thus an exterior point H is the pole of a secant h, and an interior point E (if such a point exists) is the pole of a nonsecant e. Of the points on a tangent p, one (namely P) is on the conic, and all the others are exterior. If p and q are any two distinct tangents, the point $p \cdot q$ (which is the pole of the secant PQ) is exterior.

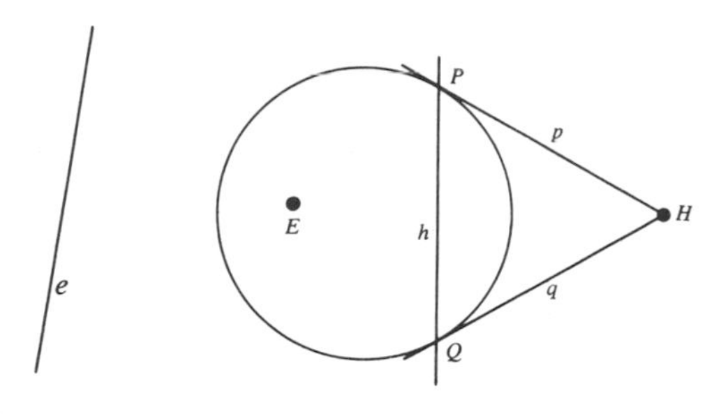

FIGURE 8.1B

Figure 8.1B may help to clarify these ideas, but we must take care not to be unduly influenced by real geometry. For instance, it would be foolish to waste any effort on trying to prove that every point on a nonsecant is exterior, or that every line through an interior point is a secant; for in some geometries these "obvious" propositions are false.

Consider the problem of drawing a secant through a given point A. If A is interior, we simply join A to any point P on the conic. (Since AP cannot be a tangent, it must be a secant.) If A is on the conic, we join it to another point on the conic. Finally, if A is exterior, we join it to each of three points on the conic. Since at most two of the lines so drawn could be tangents, at least one must be a secant.

On a secant PQ, the involution of conjugate points is $(PP)(QQ)$. Hence, by 5.41,

8.11 *Any two conjugate points on a secant PQ are harmonic conjugates with respect to P and Q.*

Conversely,

8.12 *On a secant PQ, any pair of harmonic conjugates with respect to P and Q is a pair of conjugate points with respect to the conic.*

Dually,

8.13 *Any two conjugate lines through an exterior point $p \cdot q$ are harmonic conjugates with respect to the two tangents p, q; and any pair of harmonic conjugates with respect to p and q is a pair of conjugate lines with respect to the conic.*

EXERCISES

1. Every point on a tangent is conjugate to the point of contact. Dually, the tangent itself is conjugate to any line through the point of contact.
2. The polar of any exterior point joins the points of contact of the two tangents that can be drawn through the point. Dually, the pole of a secant PQ is the point of intersection of the tangents at P and Q.
3. Is it true that every conic has exterior points?
4. Is it true that the polar of any interior point is a nonsecant?
5. If PQR is a triangle inscribed in a conic, the tangents at P, Q, R form a triangle circumscribed about the conic. These are perspective triangles. (*Hint:* Use Chasles's theorem.)
6. Any two vertices of a triangle circumscribed to a conic are separated harmonically by the point of contact of the side containing them and the point where this side meets the line joining the points of contact of the other sides. (Reference **19**, p. 140.)
7. In the spirit of Section 2.5, Exercise 5, what would be natural names for:
   (i) a conic of which the line at infinity is a nonsecant,
   (ii) a conic of which the line at infinity is a tangent,
   (iii) a conic of which the line at infinity is a secant,
   (iv) the pole of the line at infinity with respect to any conic,
   (v) the tangents of a hyperbola through its center,
   (vi) a line (other than a tangent) through the center of any conic,
   (vii) conjugate lines through the center?
8. Continuing to work in the "affine" geometry of Exercise 7, let D be the point of intersection of the tangents at any two points P and Q on a parabola, let C be the midpoint of PQ, and let S be the midpoint of CD. Then S lies on the parabola.

8.2 The Polarity Induced by a Conic

We have seen that any hyperbolic polarity determines a conic. Conversely, any conic (given as a locus of points) determines a hyperbolic polarity. A suitable construction will emerge as a by-product of the following theorem:

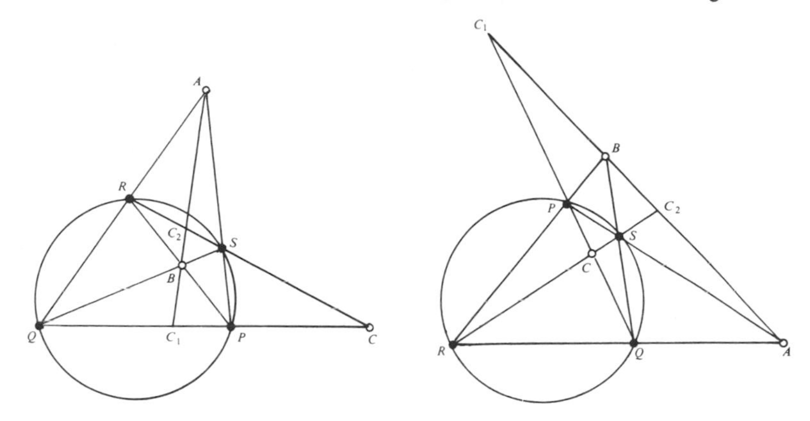

FIGURE 8.2A

8.21 *If a quadrangle is inscribed in a conic, its diagonal triangle is self-polar.*

PROOF. Let the diagonal points of the inscribed quadrangle $PQRS$ be

$$A = PS \cdot QR, \quad B = QS \cdot RP, \quad C = RS \cdot PQ,$$

as in Figure 8.2A. The line AB meets the sides PQ and RS in points C_1 and C_2 such that $\mathrm{H}(PQ, CC_1)$ and $\mathrm{H}(RS, CC_2)$. By 8.12, both C_1 and C_2 are conjugate to C. Thus the line AB, on which they lie, is the polar of C. Similarly, BC is the polar of A, and CA of B.

Hence:

8.22 *To construct the polar of a given point C, not on the conic, draw any two secants PQ and RS through C; then the polar joins the two points $QR \cdot PS$ and $RP \cdot QS$.*

In other words, we draw two secants through C to form an inscribed quadrangle with diagonal triangle ABC, and then the polar of C is AB.

The dual construction presupposes that we know the tangents from any exterior point. This presents no serious difficulty (since their points of contact lie on the polar of the given point); but the tangents are not immediately apparent, for the simple reason that we are in the habit of dealing with loci rather than envelopes. If we insist on regarding the conic as a locus, we can construct the pole of a given line as the common point of the polars of any two points on the line. Then:

8.23 *To construct the tangent at a given point P on the conic, join P to the pole of any secant through P.*

These constructions serve to justify the statement that any conic determines a hyperbolic polarity whose self-conjugate points are the points on the conic.

EXERCISES

1. Let A and B be two conjugate points with respect to a given conic. Let an arbitrary line through A meet the conic in Q and R, while BQ and BR meet the conic again in S and P, respectively. Then A, S, P are collinear.
2. If PQR is a triangle inscribed in a conic, any point A on QR (except Q or R or $p \cdot QR$) is a vertex of a self-polar triangle ABC with B on RP and C on PQ.
3. If ABC is a self-polar triangle for a given conic, any secant QR through A provides a side of an inscribed triangle PQR whose remaining sides pass through B and C, respectively.
4. A conic is transformed into itself by any harmonic homology whose center is the pole of its axis.
5. The polars of the vertices of any quadrangle $PQRS$ form a quadrilateral $pqrs$ such that the 3 diagonal points of $PQRS$ are the poles of the 3 diagonal lines of $pqrs$. (See the exercise at the end of Chapter 6.) What happens to these 3 points and 3 lines in the special case when $PQRS$ is an *inscribed* quadrangle, so that p, q, r, s are tangents?
6. There is, in general, just one conic through three given points having another given point as pole of a given line. (Reference **19**, p. 137.) [*Hint:* Let PQR be the given triangle, D and d the point and line. Let DP meet d in P'. Let S be the harmonic conjugate of P with respect to D and P'. Let ABC be the diagonal triangle of the quadrangle $PQRS$. Consider the polarity $(ABC)(Dd)$.]

8.3 Projectively Related Pencils

The following theorem, due to Franz Seydewitz (1807–1852), provides a useful connection between conjugate points and conjugate lines:

8.31 SEYDEWITZ'S THEOREM: *If a triangle is inscribed in a conic, any line conjugate to one side meets the other two sides in conjugate points.*

PROOF. Consider an inscribed triangle PQR, as in Figure 8.2A. Any line c conjugate to PQ is the polar of some point C on PQ. Let RC meet the conic again in S. By 8.21, the diagonal points of the quadrangle $PQRS$ form a self-polar triangle ABC whose side c contains the conjugate points A and B: one on QR and the other on RP.

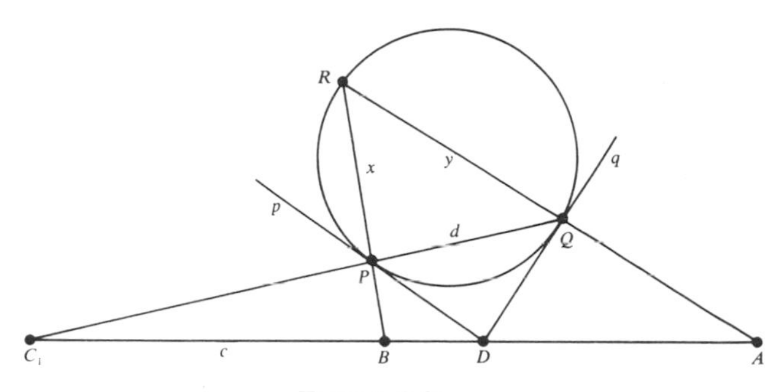

FIGURE 8.3A

We are now ready for one of the most significant properties of a conic:

8.32 STEINER'S THEOREM: *Let lines x and y join a variable point on a conic to two fixed points on the same conic; then $x \barwedge y$.*

PROOF. The tangents p and q, at the fixed points P and Q, meet in D, the pole of PQ (see Figure 8.3A). Let c be a fixed line through D (but not through P or Q), meeting x in B, and y in A. By 8.31, BA is a pair of the involution of conjugate points on c. Hence, when the point $R = x \cdot y$ varies on the conic, we have

$$x \barwedge B \barwedge A \barwedge y,$$

as desired.

The following remarks make it natural for us to include the tangents p and q as special positions for x and y. (Notice that the idea of "continuity," though intuitively helpful, is not assumed.) Writing $d = PQ$ and $C_1 = c \cdot d$, consider the possibility of using P or Q as a position for R. When R is P, y is d, A is C_1, B is the conjugate point D, and therefore x is p. Similarly, when R is Q, x is d, B is C_1, A is D, and y is q. In other words, when y is d, x is p; and when x is d, y is q.

EXERCISES

1. Dualize Seydewitz's theorem and Steiner's theorem (Figures 8.2A and 8.3A).

2. ("The Butterfly Theorem.") Let P, Q, R, S, T, U be 6 points on a conic, such that the lines PS, QR, TU all pass through a point A. Also let TU meet PR in E, and QS in B. Then

$$TEAU \barwedge TABU.$$

8.4 Conics Touching two Lines at Given Points

In the proof of 8.32, we chose an arbitrary line c through D (the pole of PQ). For any particular position of R, we can usefully take c to be a side of the diagonal triangle ABC of the inscribed quadrangle $PQRS$, where S is on RD, as in Figure 8.4A; that is, we define $C = PQ \cdot RS$ and let c be its polar AB, which passes through D since C lies on d. The point $C_1 = c \cdot d$, being the pole of the line CD, is the harmonic conjugate of C with respect to P and Q.

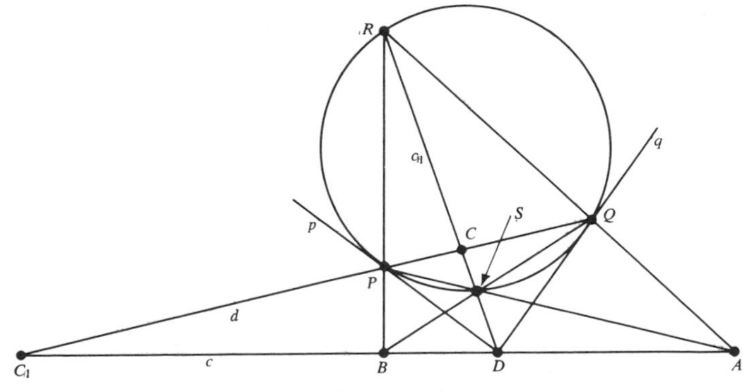

FIGURE 8.4A

If we are not given the conic, but only the points P, Q, R, D, we can still construct $C = PQ \cdot RD$ and its harmonic conjugate C_1. Then c is the line C_1D, which meets QR and RP in A and B. The conic itself can be described as the locus of self-conjugate points (and the envelope of self-conjugate lines) in the polarity $(ABC)(Pp)$, where $p = PD$. Since PD and QD are the tangents at P and Q, our conclusion may be stated as follows:

8.41 *A conic is determined when three points on it and the tangents at two of them are given.*

Retaining P, p, Q, q, but letting R vary, we obtain a "pencil" of conics touching p at P and q at Q. Such conics are said to have *double contact* (with one another). Let one of them meet a fixed line h in R and S (see Figure 8.4B). Let h meet the fixed line $d = PQ$ in C. Let c (the polar of C) meet d in C_1, and

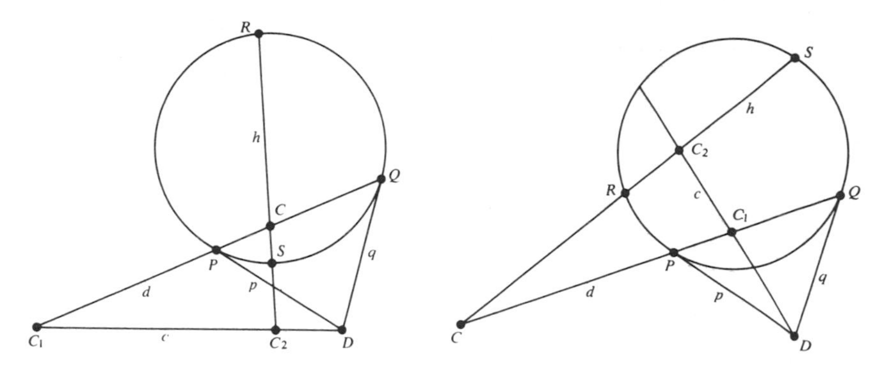

FIGURE 8.4B

h in C_2. The line c is fixed, since it joins $D = p \cdot q$ to C_1, which is the harmonic conjugate of C with respect to P and Q. In other words, the fixed point $C = d \cdot h$ has the same polar for all the conics. Thus $C_2 = c \cdot h$ is another fixed point, and RS is always a pair of the hyperbolic involution $(CC)(C_2C_2)$ on h. Hence:

8.42 *Of the conics that touch two given lines at given points, those which meet a third line (not through either of the points) do so in pairs of an involution.*

EXERCISES

1. In Figure 8.4A, RC_1 is the tangent at R. [*Hint:* Use 8.23 with R for A.]
2. Given a quadrangle $PQRD$ (as in Figure 8.4B), construct another point on the conic through R that touches PD at P and QD at Q.
3. Given three tangents to a conic, and the points of contact of two of them, construct another tangent.
4. Any two conics are related by a projective collineation and by a projective correlation. More precisely, any three distinct points on the first conic can be made to correspond to any three distinct points or tangents of the second.
5. All the conics of a double-contact pencil are transformed into themselves (each separately) by many harmonic homologies. In fact, the center of such a homology may be any point (other than P or Q) on the line PQ (Figure 8.4B). (E. P. Wigner.*)

---

* Private communication.

8.5 Steiner's Definition for a Conic

We have followed von Staudt in defining a conic by means of the self-conjugate points (and self-conjugate lines) in a hyperbolic polarity. An alternative approach is suggested by Steiner's theorem, 8.32. Could a conic be defined as the locus of the common point of corresponding lines of two projective (but not perspective) pencils? Of course, this construction would only yield a conic *locus*: there would remain the problem of deducing that its tangents join corresponding points of two projective (but not perspective) ranges. The theorem that makes such an alternative procedure possible is as follows:

8.51 *Let variable lines x and y pass through fixed points P and Q in such a way that $x \barwedge y$ but not $x \doublebarwedge y$. Then the locus of $x \cdot y$ is a conic through P and Q. If the projectivity has the effect $pdx \barwedge dqy$, where $d = PQ$, then p and q are the tangents at P and Q.*

PROOF. Since the projectivity $x \barwedge y$ is not a perspectivity, the line $d = PQ$ (Figure 8.3A) does not correspond to itself. Hence there exist lines p and q such that the projectivity relates p to d, and d to q. By 8.41, there is a unique conic touching p at P, q at Q, and passing through any other particular position of the variable point $x \cdot y$. By 8.32, this conic determines a projectivity relating all the lines through P to all the lines through Q. By the fundamental theorem, the two projectivities must coincide, since they agree for three particular positions of x and the corresponding positions of y.

EXERCISES

1. Given a triangle PQR and a point O, not on any side, what is the locus of the trilinear pole of a variable line through O? [*Hint:* In Figure 3.4A, let PQR be fixed while DE varies in a pencil. Then $A \barwedge D \doublebarwedge E \barwedge B$.]
2. Let P and Q be two fixed points on a tangent of a conic. If x is a variable line through P, and X is the (variable) pole of x, what is the locus of $x \cdot QX$? (S. Schuster.*)
3. Give an explicit determination of the locus in Exercise 2, by naming a sufficient number of special points on it.
4. Let P, Q, R, P', Q' be five points, no three collinear, and let x be a variable line through P. Define

   $N = PQ' \cdot P'Q, \quad M = RP' \cdot x, \quad L = Q'R \cdot MN, \quad R' = QL \cdot x.$

   What is the locus of R'?

* Private communication.

CHAPTER NINE

The Conic, Continued

> Had [Pascal] confined his attention to mathematics he might have enriched the subject with many remarkable discoveries. But after his early youth he devoted most of his small measure of strength to theological questions.
>
> *J. L. Coolidge* (Reference **5**, p. 89)

9.1 The Conic Touching Five Given Lines

Dualizing 8.51 (as in Figure 9.1A) we obtain

9.11 *Let points X and Y vary on fixed lines p and q in such a way that $X \barwedge Y$ but not $X \doublebarwedge Y$. Then the envelope of XY is a conic touching p and q. If the projectivity has the effect $PDX \barwedge DQY$, where $D = p \cdot q$, then P and Q are the points of contact of p and q.*

Let X_1, X_2, X_3 be three positions of X on p, and Y_1, Y_2, Y_3 the corresponding positions of Y on q, as in Figure 9.1B. By 4.12, there is a unique

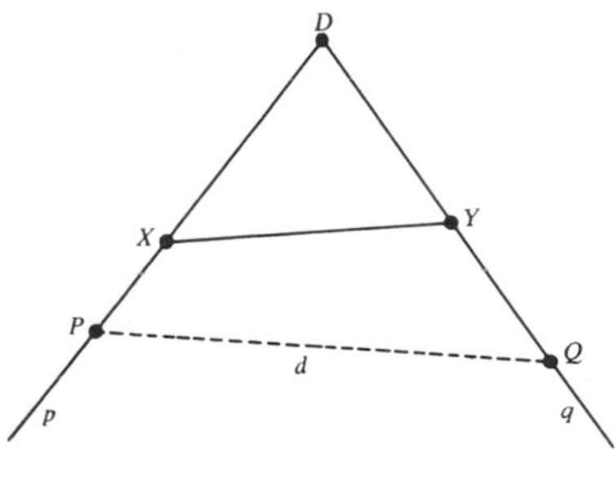

FIGURE 9.1A

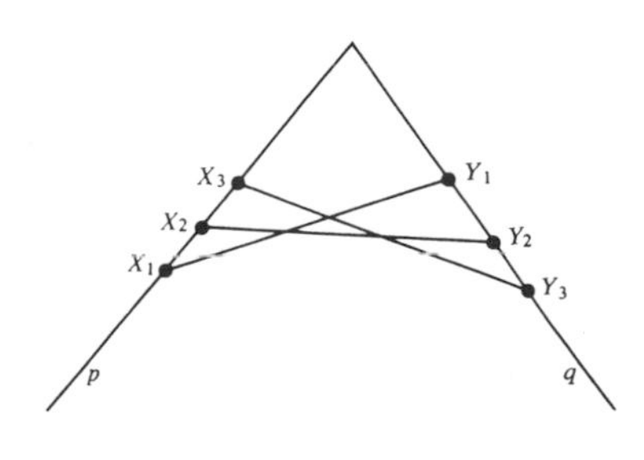

FIGURE 9.1B

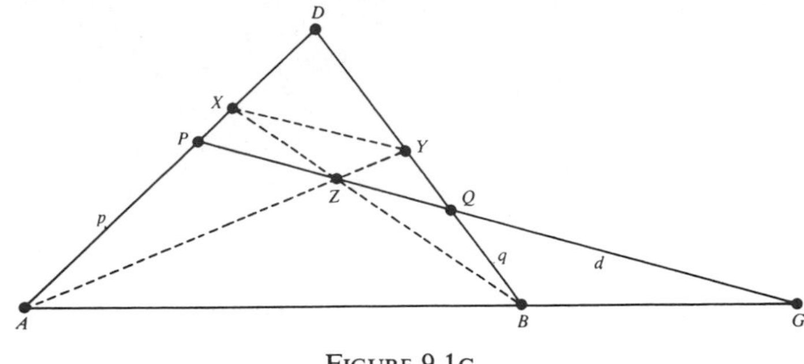

FIGURE 9.1C

projectivity $X_1X_2X_3X \barwedge Y_1Y_2Y_3Y$. By Theorem 9.11, the envelope of XY is a conic, provided no three of the five lines X_iY_i, p, q are concurrent. Conversely, if five such lines all touch a conic, any other tangent XY satisfies

$$X_1X_2X_3X \barwedge Y_1Y_2Y_3Y.$$

Hence

9.12 *Any five lines, of which no three are concurrent, determine a unique conic touching them.*

By 4.32, the line $d = PQ$ (Figure 9.1A) is the axis of the projectivity $X \barwedge Y$; that is, if A is a particular position of X and B is the corresponding position of Y (Figure 9.1C), the point $Z = AY \cdot BX$ always lies on this fixed line d. In fact, if AB meets d in G, we have an expression for the projectivity as the product of two perspectivities:

$$APDX \overset{B}{\barwedge} GPQZ \overset{A}{\barwedge} BDQY.$$

We may regard XYZ as a variable triangle whose vertices run along fixed lines p, q, d while the two sides YZ and ZX pass through fixed points A and B. We have seen that the envelope of XY is a conic touching p at P, and q at Q. More generally,

9.13 *If the vertices of a variable triangle lie on three fixed nonconcurrent lines p, q, r, while two sides pass through fixed points A and B, not collinear with $p \cdot q$, then the third side envelops a conic.*

PROOF. Let XYZ be the variable triangle, whose vertices X, Y, Z run along the fixed lines p, q, r while the sides YZ and ZX pass through points A and B (not necessarily on p or q), as in Figure 9.1D. Then

$$X \overset{B}{\barwedge} Z \overset{A}{\barwedge} Y.$$

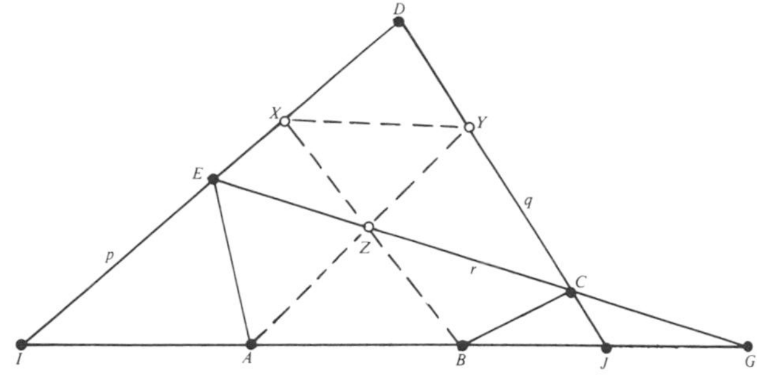

FIGURE 9.1D

Since neither r nor AB passes through $D = p \cdot q$, the projectivity $X \barwedge Y$ is not a perspectivity. By Theorem 9.11, the envelope of XY is a conic touching p and q.

In Figure 9.1D, each position for Z on r yields a corresponding position for the tangent XY. Certain special positions are particularly interesting. Defining

$$C = q \cdot r, \quad E = p \cdot r, \quad G = AB \cdot r, \quad I = AB \cdot p, \quad J = AB \cdot q,$$

we see that, when Z is E, X also is E, AY is AE, and XY also is AE. Similarly, when Z is C, Y also is C, BX is BC, and XY also is BC. Finally, when Z is G, X is I, Y is J, and XY is AB. Thus the lines AE, BC, AB, like p and q, are special positions for XY. In other words, all five sides of the pentagon $ABCDE$ are tangents of the conic. We now have the following construction for any number of tangents of the conic inscribed in a given pentagon $ABCDE$.

9.14 *Let Z be a variable point on the diagonal CE of a given pentagon $ABCDE$. Then the two points*

$$X = ZB \cdot DE, \qquad Y = ZA \cdot CD,$$

determine a line XY whose envelope is the inscribed conic.

In Figure 9.1D, we see a hexagon $ABCYXE$ whose six sides all touch a conic. The three lines AY, BX, CE, which join pairs of opposite vertices, are naturally called *diagonals* of the hexagon. Theorem 9.14 tells us that, if the diagonals of a hexagon are concurrent, the six sides all touch a conic. Conversely, if all the sides of a hexagon touch a conic, five of them can be identified with the lines DE, EA, AB, BC, CD. Since the given conic is the only one that touches these fixed lines, the sixth side must coincide with one of the lines XY for which $BX \cdot AY$ lies on CE. We have thus proved

9.15 BRIANCHON'S THEOREM: *If a hexagon is circumscribed about a conic, the three diagonals are concurrent.*

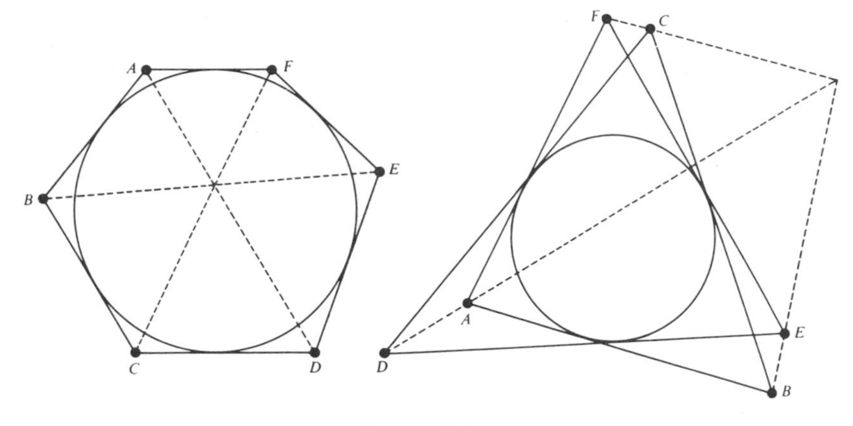

FIGURE 9.1E

Figure 9.1E illustrates this in a more natural notation: the Brianchon hexagon is $ABCDEF$ and its diagonals are AD, BE, CF.

EXERCISES

1. Apply Figure 9.1C to Exercise 3 of Section 8.4.
2. Obtain a simple construction for the point of contact of any one of five given tangents of a conic. [*Hint:* To locate the point of contact of p, in the notation of Figure 9.1D, make Y coincide with D, as in Figure 9.1F.]
3. Measure off points $X_0, X_1, \ldots, X_5$ at equal intervals along a line, and $Y_0, Y_1, \ldots, Y_5$ similarly along another line through $X_5 = Y_0$. What kind of conic do the joins $X_k Y_k$ appear to touch? Draw some more tangents.

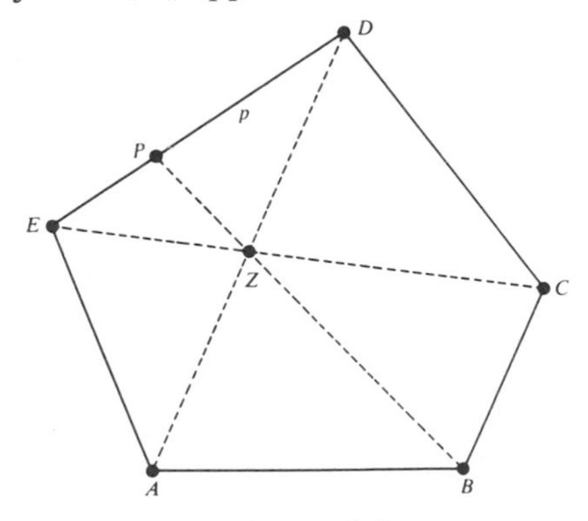

FIGURE 9.1F

9.2 The Conic Through Five Given Points

Dualizing 9.12 (as in Figure 9.2A), we obtain

9.21 *Any five points, of which no three are collinear, determine a unique conic through them.*

The dual of 9.13 was discovered independently by William Braikenridge and Colin MacLaurin, about 1733:

9.22 *If the sides of a variable triangle pass through three fixed non-collinear points P, Q, R, while two vertices lie on fixed lines a and b, not concurrent with PQ, then the third vertex describes a conic.*

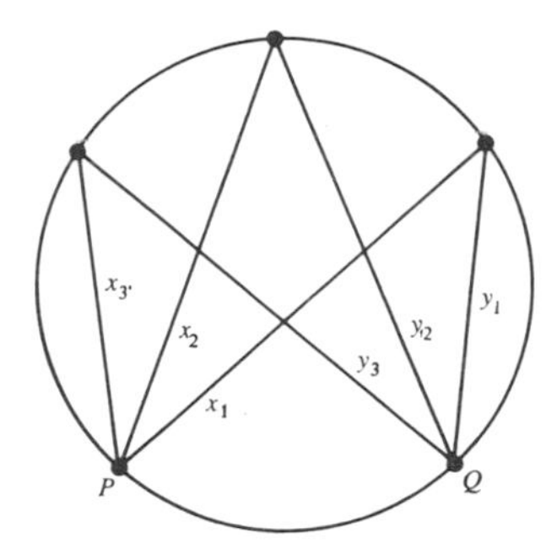

FIGURE 9.2A

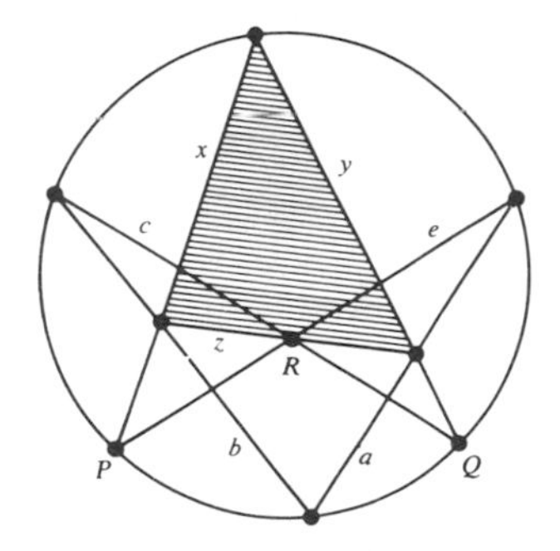

FIGURE 9.2B

This enables us (as in Figure 9.2B, where the variable triangle is shaded) to locate any number of points on the conic through five given points.

The dual of 9.15 is the still more famous

9.23 PASCAL'S THEOREM: *If a hexagon is inscribed in a conic, the three pairs of opposite sides meet in collinear points.*

In Figure 9.2C, the hexagon is *abcdef* and the three collinear points are $a \cdot d$, $b \cdot e$, $c \cdot f$. We have obtained Pascal's theorem by dualizing Brianchon's. Historically, this procedure was reversed: C. J. Brianchon (1760–1854) obtained his theorem by dualizing Pascal's, at a time when the principle of duality was just beginning to be recognized. Pascal's own proof (for a hexagon inscribed in a circle) was seen and praised by G. W. Leibniz (1646–1716) when he visited Paris, but afterwards it was lost. All that remains of Pascal's relevant work is a brief *Essay pour les coniques* (1640), in which the theorem is stated as follows:

Si dans le plan *MSQ* du point *M* partent les deux droites *MK*, *MV*, & du point

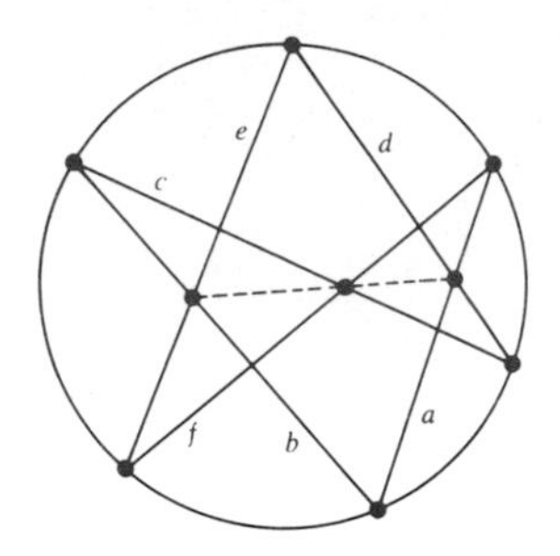

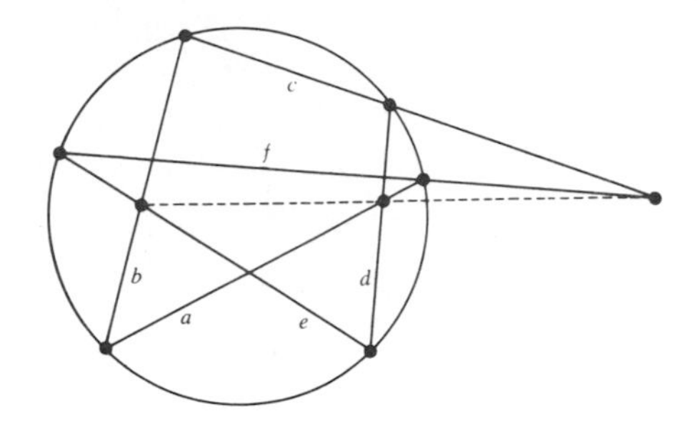

FIGURE 9.2C

S partent les deux droites *SK*, *SV* · · · & par les points *K*, *V* passe la circonférence d'un cercle coupante les droites *MV*, *MK*,* *SV*, *SK* ès pointes *O*, *P*, *Q*, *N*: je dis que les droites *MS*, *NO*, *PQ* sont de mesme ordre.

EXERCISES

1. Given five points (no three collinear), construct the tangent at each point to the conic through all of them.
2. Given a quadrangle *PQRS* and a line *s* through *S* (but not through any other vertex), construct another point on the conic through *P*, *Q*, *R* that touches *s* at *S*.
3. Given six points on a conic, in how many ways can they be regarded as the vertices of a Pascal hexagon?
4. Name the hexagon in Pascal's own notation.
5. Try to reconstruct Pascal's lost proof (using only the methods that would have been available in his time).

---

* *Œuvres de Blaise Pascal*, edited by L. Brunschvicg and P. Boutroux, **1** (Libraire Hachette: Paris, 1908), p. 252. Pascal actually wrote *MP* for *MK*, but this was obviously a slip. By "de mesme ordre" he meant "in the same pencil" or, in the terminology of modern projective geometry, "concurrent." Compare "d'une mesme ordonnance" in the passage of Desargues that we quoted on page 3. Pascal was the first person who properly appreciated the work of Desargues. The complete statement may be translated as follows:

If, in the plane *MSQ*, two lines *MK* and *MV* are drawn through *M*, and two lines *SK*, *SV* through *S*, and if a circle through *K* and *V* meets the four lines *MV*, *MK*, *SV*, *SK* in points *O*, *P*, *Q*, *N*, then the three lines *MS*, *NO*, *PQ* belong to a pencil.

Although nobody knows just how Pascal proved this property of a circle, there is no possible doubt about how he deduced the analogous property of the general conic. He joined the circle and lines to a point outside the plane, obtaining a cone and planes; then he took the section of this solid figure by an arbitrary plane.

9.3 Conics Through Four Given Points

Desargues not only invented the word *involution* (in its geometrical sense) but also showed how the pairs of points belonging to an involution on a line arise from a "pencil" of conics through four points. This is his "involution theorem," which is even more remarkable than his "two triangle theorem."

9.31 DESARGUES'S INVOLUTION THEOREM: *Of the conics that can be drawn through the vertices of a given quadrangle, those which meet a given line (not through a vertex) do so in pairs of an involution.*

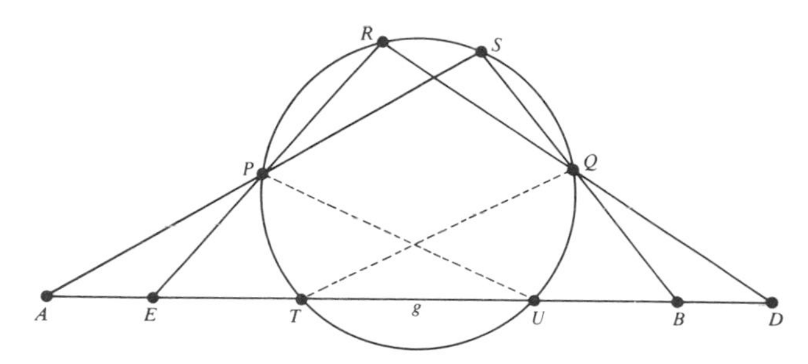

FIGURE 9.3A

PROOF. Let $PQRS$ be the given quadrangle, and g the given line, meeting the sides PS, QS, QR, PR in A, B, D, E, and any one of the conics in T and U (see Figure 9.3A). By regarding S, R, T, U as four positions of a variable point on this conic, we see from 8.32 that the four lines joining them to P are projectively related to the four lines joining them to Q. Hence

$$AETU \barwedge BDTU.$$

Since, by Theorem 1.63, $BDTU \barwedge DBUT$, it follows that

$$AETU \barwedge DBUT.$$

Hence TU is a pair of the involution $(AD)(BE)$. Since this involution depends only on the quadrangle, all those conics of the pencil which intersect g (or touch g) determine pairs (or invariant points) of the same involution.

Referring to Figure 9.3A again, we observe that, when S and Q coincide, the line SQ (which determines B) is replaced by the tangent at Q. Everything else remains. Hence:

9.32 *Of the conics that can be drawn to touch a given line at a given point while passing also through two other given points, those which meet another given line (not through any of the three given points) do so in pairs of an involution.*

Similarly, by letting R and P coincide, we obtain an alternative proof for Theorem 8.42.

EXERCISES

1. Given five points P, Q, R, S, T, no three collinear, and a line g through P, construct the second common point of the conic $PQRST$ and the line g.
2. A given line touches at most two of the conics through the vertices of a given quadrangle.
3. Let P, Q, R, S, T, U be 6 points on a conic, such that the lines PS, QR, TU all pass through a point A. Also let TU meet PR in E, and QS in B. Then EB is a pair of the involution $(AA)(TU)$.
   Can this be deduced directly from Exercise 2 of Section 8.3?
4. Let P, Q, R, S be four points on a conic, and t the tangent at a fifth point. If no diagonal point of the quadrangle $PQRS$ lies on t, there is another conic also passing through P, Q, R, S and touching t.

9.4 Two Self-Polar Triangles

Combining 9.31 with 7.22, we see that the involution determined on g (Figure 9.3A) by the quadrangle $PQRS$ is not only the Desargues involution determined by conics through P, Q, R, S but also the involution of conjugate points on g for the polarity $(PQR)(Sg)$. Hence:

9.41 *If two triangles have six distinct vertices, all lying on a conic, there is a polarity for which both triangles are self-polar.*

And conversely (Figure 9.4A),

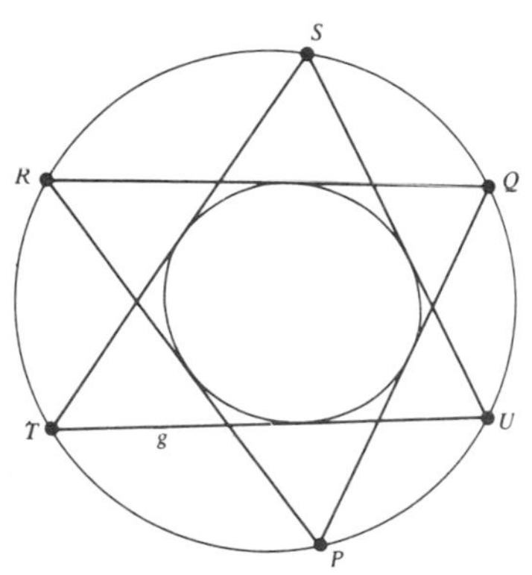

FIGURE 9.4A

9.42 *If two triangles, with no vertex of either on a side of the other, are self-polar for a given polarity, their six vertices lie on a conic and their six sides touch another conic.*

EXERCISES

1. How many polarities can be expected to arise in the manner of 9.41 from six given points on a conic?
2. If two triangles have six distinct vertices, all lying on a conic, their six sides touch another conic.
3. If two conics are so situated that there is a triangle inscribed in one and circumscribed about the other, then every secant of the former conic that is a tangent of the latter can be used as a side of such an inscribed-circumscribed triangle.
4. Let P, Q, R, S, T be five points, no three collinear. Then the six points

$$A = QR \cdot PS, \qquad B = RP \cdot QS, \qquad C = PQ \cdot RS,$$
$$A' = QR \cdot PT, \qquad B' = RP \cdot QT, \qquad C' = PQ \cdot RT$$

all lie on a conic. (S. Schuster.)

9.5 Degenerate Conics

For some purposes it is convenient to admit, as *degenerate* conics, a pair of lines (regarded as a locus) or a pair of points (regarded as an envelope: the set of all lines through one or both). Visibly (Figure 9.5A) a hyperbola may differ as little as we please from a pair of lines (its asymptotes), and the set of tangents of a very thin ellipse is hardly distinguishable from the lines through one or other of two fixed points.

By omitting the phrase "but not $x \barwedge y$" from the statement of Steiner's construction 8.51, we could allow the locus to consist of two lines: the axis of the perspectivity $x \barwedge y$, and the line PQ (any point of which is joined to P

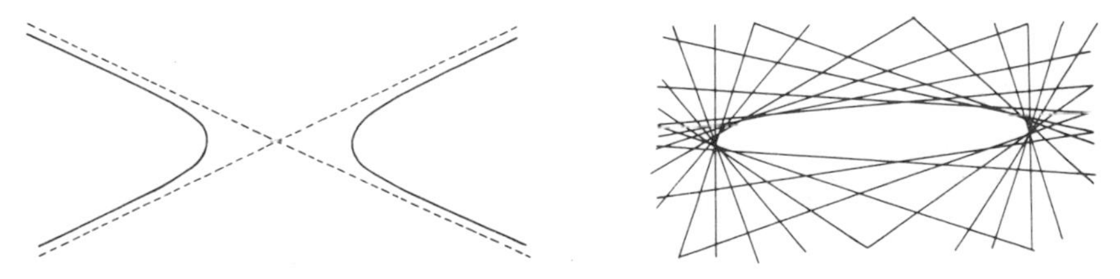

FIGURE 9.5A

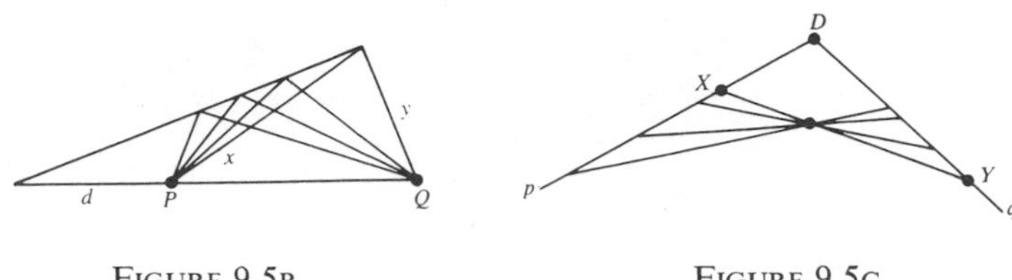

FIGURE 9.5B FIGURE 9.5C

and Q by "corresponding lines" of the two pencils, namely by the invariant line PQ itself, as in Figure 9.5B).

Dually (Figure 9.5C), when the points P and Q of Figure 8.5A coincide with D, we have a degenerate conic envelope consisting of two points, regarded as two pencils: the various positions of the line XY when X and Y are distinct, and the pencil of lines through D.

In the same spirit we can say that a conic is determined by five points, no *four* collinear, or by five lines, no four concurrent.

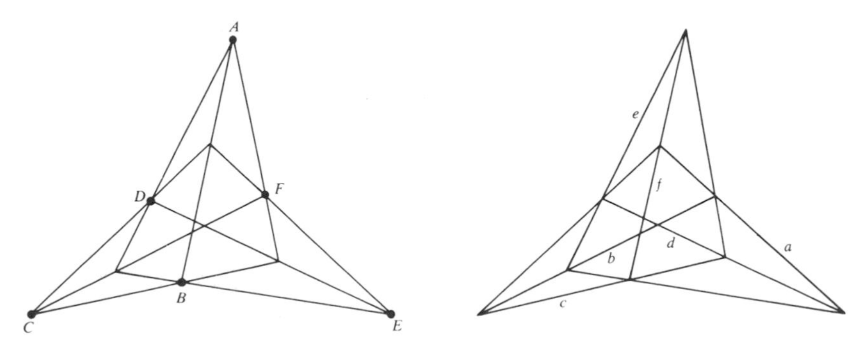

FIGURE 9.5D

The degenerate forms of Brianchon's theorem (Figure 9.1D) and Pascal's theorem (Figure 9.2C) are as follows:

If AB, CD, EF are concurrent and DE, FA, BC are concurrent, then AD, BE, CF are concurrent.

If $a \cdot b$, $c \cdot d$, $e \cdot f$ are collinear and $d \cdot e$, $f \cdot a$, $b \cdot c$ are collinear, then $a \cdot d$, $b \cdot e$, $c \cdot f$ are collinear.

Comparing Figure 9.5D with Figure 4.4A, we see that both these statements are equivalent to Pappus's theorem, 4.41.

EXERCISES

1. What kind of "polarity" is induced by a degenerate conic?
2. What happens to Exercise 4 of Section 9.3 if we omit the words "If no diagonal point of the quadrangle $PQRS$ lies on t?"

CHAPTER TEN

A Finite Projective Plane

Our Geometry is an abstract Geometry. The reasoning could be followed by a disembodied spirit who had no idea of a physical point; just as a man blind from birth could understand the Electromagnetic Theory of Light.

H. G. Forder (1889–   )
(Reference **9**, p. 43)

10.1 The Idea of a Finite Geometry

The above words of Forder emphasize the fact that our primitive concepts are defined solely by their properties as described in the axioms. This fact is most readily appreciated when we abandon the "intuitive" idea that the number of points is infinite. We shall find that all our theorems remain valid (although the figures are somewhat misleading) when there are only 6 points on each line, and 31 points in the plane.

In 1892, Fano described an n-dimensional geometry in which the number of points on each line is $p + 1$ for a fixed prime p. In 1906, O. Veblen and W. H. Bussey gave this finite Projective Geometry the name $PG(n, p)$ and extended it to $PG(n, q)$, where $q = p^k$, p is prime, and k is any positive integer. (For instance, q may be 5, 7, or 9, but cannot be 6.)

Without realizing the necessity for restricting the possible values of q to primes and their powers, von Staudt obtained the following numerical results in 1856. Since any range or pencil can be related to any other by a sequence of elementary correspondences, the number of points on a line must be the same for all lines, and the same as the number of lines in a pencil (that is, lying in a plane and passing through a point) or the number of planes through a line in three-dimensional space. Let us agree to call this number $q + 1$. In a plane, any one point is joined to the remaining points by a pencil which consists of $q + 1$ lines, each containing the one point and q others. Hence the plane contains

$$q(q+1) + 1 = q^2 + q + 1$$

points and (dually) the same number of lines. In space, any line l is joined to the points outside l by $q + 1$ planes, each containing the $q + 1$ points on l and q^2 others. Hence the whole space contains

$$(q+1)(q^2+1) = q^3 + q^2 + q + 1$$

points and (dually) the same number of planes.

The general formula for the number of points in $PG(n, q)$ is

$$q^n + q^{n-1} + \cdots + q + 1 = \frac{q^{n+1} - 1}{q - 1}.$$

It was proved by J. Singer (Trans. Amer. Math. Soc. **43** (1938), pp. 377–385) that every geometry of this kind can be represented by a combinatorial scheme such as the one exhibited on page 94 for the special case $PG(2, 5)$.

EXERCISES

1. In $PG(3, q)$ there are $q + 1$ points on each line, how many lines (or planes) through each point? How many lines in the whole space? [*Hint:* Every two of the $q^3 + q^2 + q + 1$ points determine a line, but each line is determined equally well by any two of its $q + 1$ points.]
2. How many triangles occur in $PG(2, q)$?
3. In the notation of Section 3.2, $PG(2, q)$ is a configuration n_d. Express n and d in terms of q.

10.2 A Combinatorial Scheme for $PG(2, 5)$

In accordance with the general formula, the finite projective plane $PG(2, 5)$ has 6 points on each line, 6 lines through each point,

$$5^2 + 5 + 1 = \frac{5^3 - 1}{5 - 1} = 31$$

points altogether, and of course also 31 lines. The appropriate scheme uses symbols $P_0, P_1, \ldots, P_{30}$ for the 31 points, and $l_0, l_1, \ldots, l_{30}$ for the 31 lines, with a table (page 94) telling us which are the 6 points on each line and which are the 6 lines through each point.

For good measure, this table gives every relation of incidence twice: each column tells us which points lie on a line and also which lines pass through a point; e.g., the last column says that the line l_0 contains the six points

$$P_0, P_1, P_3, P_8, P_{12}, P_{18}$$

and that the point P_0 belongs to the six lines

$$l_0, l_1, l_3, l_8, l_{12}, l_{18}.$$

Thus the notation exhibits a polarity $P_r \longleftrightarrow l_r$. Marshall Hall (*Cyclic projective planes*, Duke Math. J. **14** (1947), pp. 1079–1090) has proved that such a polarity always occurs. (Our l_r is his m_{-r}.) By regarding the subscripts as residues modulo 31, so that $r + 31$ has the same significance as r itself, we can condense the whole table into the simple statement that the point P_r and line l_s are incident if and only if

10.21 $$r + s \equiv 0, 1, 3, 8, 12, \text{ or } 18 \pmod{31}.$$

The "congruence" $a \equiv b \pmod{n}$ is a convenient abbreviation for the statement that a and b leave the same remainder (or "residue") when divided by n. The residues 0, 1, 3, 8, 12, 18 (mod 31) are said to form a "perfect difference set" because every possible residue except zero (namely, 1, 2, 3, . . ., 30) is uniquely expressible as the difference between two of these special residues:

$$1 \equiv 1 - 0, \quad 2 \equiv 3 - 1, \quad 3 \equiv 3 - 0, \quad 4 \equiv 12 - 8, \cdots,$$
$$13 \equiv 0 - 18, \ldots, 30 \equiv 0 - 1.$$

The impossibility of a $PG(2, 6)$ is related in a subtle manner to the impossibility of solving Euler's famous problem of the 36 officers (Reference **2**, p. 190).

EXERCISES

1. Set up a table of differences (mod 31) of the residues 0, 1, 3, 8, 12, 18, analogous to the following table of differences (mod 13) of 0, 1, 3, 9:

	0	1	3	9
0	0	1	3	9
1	12	0	2	8
3	10	11	0	6
9	4	5	7	0

Table of possible values of s, given r, such that P_r and l_s (or l_r and P_s) are incident

r	30	29	28	27	26	25	24	23	22	21	20	19	18	17	16	15	14	13	12	11	10	9	8	7	6	5	4	3	2	1	0
	1	2	3	4	5	6	7	8	9	10	11	12	13	14	15	16	17	18	19	20	21	22	23	24	25	26	27	28	29	30	0
	2	3	4	5	6	7	8	9	10	11	12	13	14	15	16	17	18	19	20	21	22	23	24	25	26	27	28	29	30	0	1
s	4	5	6	7	8	9	10	11	12	13	14	15	16	17	18	19	20	21	22	23	24	25	26	27	28	29	30	0	1	2	3
	9	10	11	12	13	14	15	16	17	18	19	20	21	22	23	24	25	26	27	28	29	30	0	1	2	3	4	5	6	7	8
	13	14	15	16	17	18	19	20	21	22	23	24	25	26	27	28	29	30	0	1	2	3	4	5	6	7	8	9	10	11	12
	19	20	21	22	23	24	25	26	27	28	29	30	0	1	2	3	4	5	6	7	8	9	10	11	12	13	14	15	16	17	18

2. Set up an incidence table for $PG(2, 3)$, assuming that P_r and l_s are incident if and only if

$$r + s \equiv 0,\ 1,\ 3, \text{ or } 9 \pmod{13}.$$

10.3 Verifying the Axioms

The discussion on pages 25 and 39 indicates that the following five axioms suffice for the development of two-dimensional projective geometry:

AXIOM **2.13** *Any two distinct points are incident with just one line.*

AXIOM **3.11** *Any two lines are incident with at least one point.*

AXIOM **3.12** *There exist four points of which no three are collinear.*

AXIOM **2.17** *The three diagonal points of a quadrangle are never collinear.*

AXIOM **2.18** *If a projectivity leaves invariant each of three distinct points on a line, it leaves invariant every point on the line.*

The fact that this is a logically consistent geometry can be established by verifying all the axioms in one special case, such as $PG(2, 5)$. To verify Axioms 2.13 and 3.11, we observe that any two residues are found together in just one column of the table, and that any two columns contain just one common number. For Axiom 3.12, we can cite $P_0P_1P_2P_5$. To check Axiom 2.17 for every complete quadrangle (or rather, for every one having P_0 for a vertex) is possible but tedious, so let us be content to take a single instance: the diagonal points of $P_0P_1P_2P_5$ are

$$l_0 \cdot l_{29} = P_3, \quad l_1 \cdot l_7 = P_{11}, \quad l_3 \cdot l_{30} = P_9.$$

Axiom 2.18 is superseded by Theorem 3.51 because a harmonic net fills the whole line. In fact, the harmonic net $\mathrm{R}(P_0P_1P_{18})$ contains the harmonic sequence $P_0P_1P_3P_{12}P_8 \cdots$. To verify this, we use the procedure suggested by Figure 3.5A, taking A, B, M, P, Q to be $P_0, P_1, P_{18}, P_5, P_{30}$, so that $C = P_3$, $D = P_{12}$, $E = P_8$, $F = P_0 = A$. Since there are only six points on the line, the sequence is inevitably periodic: the five points

$$P_0,\ P_1,\ P_3,\ P_{12},\ P_8$$

are repeated cyclically for ever. Instead of taking P and Q to be P_5 and P_{30}, we could just as well have taken them to be any other pair of points on l_{13} or l_{14} or l_{16} or l_{21} or l_{25} (these being, with l_0, the lines through P_{18}); we would still have obtained the same harmonic sequence.

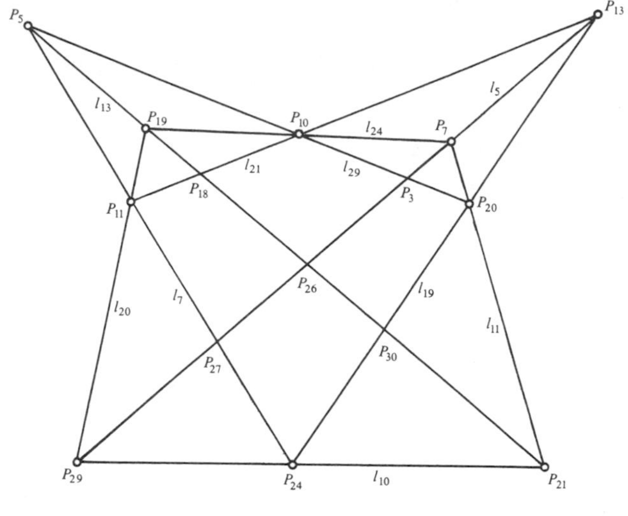

FIGURE 10.3A

EXERCISES

1. Set up an incidence table for $PG(2, 2)$, assuming that P_r and l_s are incident if and only if

$$r + s \equiv 0, 1, \text{ or } 3 \pmod 7).$$

Verify that this "geometry" satisfies all our two-dimensional axioms *except* Axiom 2.17.

2. Verify Desargues's theorem as applied to the triangles $P_{11}P_{10}P_{19}$ and $P_{24}P_{20}P_{21}$, which are perspective from the point P_5 and from the line l_5. (Figure 10.3A is the appropriate version of Figure 2.3A.)

10.4 Involutions

Turning to Figure 2.4A, we observe that the sections of the quadrangles $P_4P_5P_6P_9$, $P_{14}P_{15}P_{16}P_{19}$, $P_9P_{10}P_{11}P_{14}$ by the line l_0 yield the quadrangular and harmonic relations

$$(P_1P_8)(P_0P_3)(P_{18}P_{12}), \quad (P_{12}P_{18})(P_8P_0)(P_3P_1), \quad \mathsf{H}(P_{12}P_{18}, P_3P_8).$$

The fundamental theorem 4.11 shows that every projectivity on l_0 is expressible in the form

$$P_0P_1P_3 \barwedge P_iP_jP_k,$$

where i, j, k are any three distinct numbers selected from 0, 1, 3, 8, 12, 18. Hence there are just $6 \cdot 5 \cdot 4 = 120$ projectivities (including the identity). Of these, as we shall see, 25 are involutions: 15 hyperbolic and 10 elliptic. In fact, if i and j are any two of the six numbers, there is a hyperbolic involution $(P_iP_i)(P_jP_j)$ which interchanges the remaining four numbers in pairs in a definite way. The other two possible ways of pairing those four numbers must each determine an elliptic involution which interchanges P_i and P_j. For instance, the hyperbolic involution $(P_{12}P_{12})(P_{18}P_{18})$, interchanging P_3 and P_8, must also interchange P_0 and P_1, and is expressible as $(P_0P_1)(P_3P_8)$; but both the involutions

$$(P_1P_8)(P_0P_3), \qquad (P_0P_8)(P_1P_3)$$

interchange P_{12} and P_{18}, and are therefore elliptic.

EXERCISES

1. In $PG(2, 3)$, how many projectivities are there on a line? How many of them are involutions? How many of the involutions are elliptic?
2. In $PG(2, q)$, the $q^3 - q$ projectivities on a line include just q^2 involutions: $q(q + 1)/2$ hyperbolic and $q(q - 1)/2$ elliptic.*
3. In $PG(2, 3)$, the four points on a line form a harmonic set in every possible order. (G. Fano.†)
4. In $PG(2, 5)$, any four distinct points on a line form a harmonic set in a suitable order. In fact, $\mathsf{H}(AB, CD)$ if and only if the involution $(AB)(CD)$ is hyperbolic. In other words, each of the fifteen pairs of points on the line induces a separation of the six into three mutually harmonic pairs. (W.L. Edge, "31-*point geometry*," Math. Gazette, **39** (1955), p. 114, section 3.)

10.5 Collineations and Correlations

By 6.22 and 6.42, every projective collineation or projective correlation is determined by its effect on a particular quadrangle, such as $P_0P_1P_2P_5$. The

* This problem, for q a prime, was solved more than a hundred years ago by J. A. Serret, *Cours d'algèbre supérieure*, Tome 2, 3rd ed. (Gauthier-Villars: Paris, 1866), p. 355, or 5th ed. (Paris, 1885), pp. 381–382.

† Giornale di Matematiche **30** (1892), p. 116.

collineation may transform P_0 into any one of the 31 points, and P_1 into any one of the remaining 30. It may transform P_2 into any one of the $31 - 6 = 25$ points not collinear with the first two. The number of points that lie on at least one side of a given triangle is evidently $3 + (3 \cdot 4) = 15$; therefore the number *not* on any side is 16. Hence $PG(2, 5)$ admits altogether

$$31 \cdot 30 \cdot 25 \cdot 16 = 372000$$

projective collineations, and the same number of projective correlations.

Of the 372000 projective collineations, 775 are of period 2 (see 6.32). For, by 6.31, the number of harmonic homologies is

$$31 \cdot 25 = 775.$$

Apart from the identity, the two most obvious collineations are $P_r \to P_{5r}$ (of period 3, since $5^3 \equiv 1 \pmod{31}$) and $P_r \to P_{r+1}$ (of period 31). The criterion 6.11 assures us that they are projective. In fact, the corresponding ranges of the former on P_0P_1 and P_0P_5 are related by the perspectivity

$$P_0P_1P_3P_8P_{12}P_{18} \overset{P_{11}}{\doublebarwedge} P_0P_5P_{15}P_9P_{29}P_{28}$$

and the corresponding ranges of the latter on P_0P_1 and P_1P_2 are related by a projectivity with axis P_0P_2:

$$P_0P_1P_3P_8P_{12}P_{18} \overset{P_9}{\doublebarwedge} P_0P_2P_{30}P_{11}P_{17}P_7 \overset{P_8}{\doublebarwedge} P_1P_2P_4P_9P_{13}P_{19}.$$

EXERCISES

1. Express the collineation $P_r \to P_{5r}$ as a transformation of lines. What happens to the incidence condition (10.21)?
2. How many projective collineations exist in $PG(2, 3)$? How many of them are of period 2?
3. In $PG(2, q)$, how many points are left invariant by

   (i) an elation, (ii) a homology?

10.6 Conics

The most obvious correlation is, of course, $P_r \to l_r$. To verify that it is projective, we may use 6.41 in the form

$$P_1P_2P_4P_9P_{13}P_{19} \overset{P_8}{\doublebarwedge} P_0P_{29}P_{28}P_9P_5P_{15} \barwedge l_1l_2l_4l_9l_{13}l_{19}.$$

Being of period 2, it is a polarity. Since P_0 lies on l_0, it is a hyperbolic polarity, and determines a conic. By 8.51 (Steiner's construction), we see that the number of points on a conic (in any finite projective plane) is equal to the number of lines through a point, in the present case 6. By inspecting the incidence table, or by halving the residues 0, 1, 3, 8, 12, 18 (mod 31), we see that the conic determined by the polarity $P_r \leftrightarrow l_r$ consists of the 6 points and 6 lines

$$P_0P_4P_6P_9P_{16}P_{17}, \qquad l_0l_4l_6l_9l_{16}l_{17}.$$

The 6 lines are the tangents, By joining the 6 points in pairs, we obtain the $\binom{6}{2} = 15$ secants

$$l_1 = P_0P_{17},\ l_2 = P_6P_{16},\ l_3 = P_0P_9.\ l_8 = P_0P_4,\ l_{12} = P_0P_6,$$

$$l_{14} = P_4P_{17},\ l_{15} = P_{16}P_{17},\ l_{18} = P_0P_{16},\ l_{22} = P_9P_{17},\ l_{23} = P_9P_{16},$$

$$l_{25} = P_6P_9,\ l_{26} = P_6P_{17},\ l_{27} = P_4P_{16},\ l_{28} = P_4P_6,\ l_{30} = P_4P_9.$$

It follows (see Figure 10.3A) that the remaining 10 lines

$$l_5,\ l_7,\ l_{10},\ l_{11},\ l_{13},\ l_{19},\ l_{20},\ l_{21},\ l_{24},\ l_{29}$$

are nonsecants, each containing an elliptic involution of conjugate points.

Any two conjugate points on a secant or nonsecant determine a self-polar triangle. For instance, the secant l_1, containing the hyperbolic involution $(P_0P_0)(P_{17}P_{17})$ or $(P_2P_{30})(P_7P_{11})$, is a common side of the two self-polar triangles $P_1P_2P_{30}$, $P_1P_7P_{11}$. These two triangles are of different types: of the former, all three sides l_1, l_2, l_{30} are secants; but the sides l_7 and l_{11} of the latter are nonsecants. We may conveniently speak of triangles of the *first* type and *second* type, respectively. Since each of the 15 secants belongs to one self-polar triangle of either type, there are altogether 5 triangles of the first type and 15 of the second. (These properties of a conic are amusingly different from what happens in *real* geometry, where the sides of a self-polar triangle always consist of two secants and one nonsecant.)

There are, of course, many ways to express a given polarity by a symbol of the form $(ABC)(Pp)$; for example, the polarity $P_r \leftrightarrow l_r$ is $(P_1P_2P_{30})(P_3l_3)$ or $(P_1P_7P_{11})(P_3l_3)$ or $(P_1P_7P_{11})(P_4l_4)$. Such symbols will enable us to find the total number of polarities.

If ABC is given, there are 16 possible choices for P (not on any side) and 16 possible choices for p (not through a vertex), making $16^2 = 256$ available symbols $(ABC)(Pp)$ for polarities in which ABC is self-polar. Since each of the 16 lines contains 3 of the 16 points, just 48 of the 256 symbols have P lying on p, as in the case of $(P_1P_7P_{11})(P_4l_4)$.

When the self-polar triangle is of the first type (with every side a secant) all the six points on the conic are on sides of the triangle, P never lies on p, and each hyperbolic polarity (with ABC of this type) is named 16 times by a symbol $(ABC)(Pp)$ with P *not* on p. When only one side is a secant, 2 of the 6 points are on this side and the remaining 4 are among the 16; therefore each hyperbolic polarity (with ABC of the second type) is named 4 times with P on p and 12 times with P not on p. Conversely, if P lies on p, ABC can only be of the second type; therefore the number of such hyperbolic polarities (each accounting for 4 of the 48 symbols) is 12. Since each hyperbolic polarity (or conic) has 5 self-polar triangles of the first type and 15 of the second, the number of hyperbolic polarities in which a given triangle ABC is of the first type is one-third of 12, that is, 4. The total number of symbols $(ABC)(Pp)$ that denote hyperbolic polarities is thus

$$48 + 16 \cdot 4 + 12 \cdot 12 = 256.$$

Since we have accounted for all the available symbols,

There are no elliptic polarities in $PG(2, 5)$.*

The total number of triangles in $PG(2, 5)$ can be found as follows. There are 31 choices for the first vertex, 30 for the second, and $31 - 6 = 25$ for the third; but the three vertices can be permuted in $3! = 6$ ways. Hence the number is

$$\frac{31 \cdot 30 \cdot 25}{6} = 31 \cdot 125 = 3875.$$

We can easily deduce the number of conics. Each conic has 5 self-polar triangles of the first type, and each triangle plays this role for 4 conics; therefore the number of conics is†

$$\frac{31 \cdot 125 \cdot 4}{5} = 3100.$$

As an instance of 7.71, we observe that the collineation $P_r \rightarrow P_{r+1}$ (or $l_r \rightarrow l_{r-1}$) is the product of the polarities $P_r \leftrightarrow l_r$ and $P_r \leftrightarrow l_{r-1}$, which may be expressed as

$$(P_3P_5P_{29})(P_4l_4) \quad \text{and} \quad (P_4P_5P_{28})(P_3l_2).$$

EXERCISES

1. In $PG(2, 5)$, express the collineation $P_r \rightarrow P_{5r}$ (or $l_s \rightarrow l_{5s+3}$) as the product of two polarities.
2. Derive the number of conics in $PG(2, 5)$ from the number of sets of 5 points, no 3 collinear.

* In other words, *every polarity is hyperbolic*. This is true not only in complex geometry and in $PG(2, 5)$ but in $PG(2, q)$ for every $q = p^k$. See P. Scherk, Can. Math. Bull., **2** (1959), pp. 45–46, or Segre (Reference **15**, pp. 266–268).

† Analogous reasoning shows that the number of conics in $PG(2, q)$ is $q^5 - q^2$. See B. Segre, *Le geometrie di Galois*, Ann. Matematica (4), **48** (1959), pp. 1–96, especially p. 4.

3. In $PG(2, 5)$, how many conics can be drawn through the vertices of a given triangle? [*Hint:* Use 9.21.] How many triangles can be inscribed in a given conic? (These results provide another method for determining the total number of conics.)

4. How many conics exist in $PG(2, 3)$? In what sense does 9.21 remain valid when there are only 4 points on a conic?

5. Does a conic in $PG(2, 3)$ admit a self-polar triangle all of whose sides are nonsecants?

6. In any projective plane, let us say that a point Q is *accessible* from a point P if it is the harmonic conjugate of P with respect to some pair of distinct points which are conjugate (to each other) in a given polarity. Then P is accessible from itself; if Q is accessible from P, P is accessible from Q; if Q is accessible from R, and R from P, Q is accessible from P. In other words, the relation of accessibility is *reflexive, symmetric, and transitive.*

7. When the notion of accessibility (Exercise 6) is applied to a hyperbolic polarity, which points are accessible from:
   (i) a point on the conic,
   (ii) an exterior point,
   (iii) an interior point?

8. If the notion of accessibility could be applied to an elliptic polarity in $PG(2, q)$, how many points would be accessible from a given point P:
   (i) on a line through P,
   (ii) in the whole plane?
   (F. Bachmann, *Aufbau der Geometrie aus dem Spiegelungsbegriff*, Springer Verlag: Berlin, 1959, pp. 123–124.)

9. There are no elliptic polarities in $PG(2, q)$.

CHAPTER ELEVEN

Parallelism

Projective geometry was historically developed as a part of Euclidean geometry.... Von Staudt, for instance, still needed the parallel axiom in his study of the foundations of projective geometry. Klein, in his work on non-Euclidean geometry, established the independence of projective geometry from the theory of parallels (in 1871). This opened the possibility of an independent foundation for projective geometry. The pioneer work was done by Moritz Pasch (in 1882).

D. J. Struik (1894–)
(Reference **17**, p. 72)

11.1 Is the Circle a Conic?

The attentive reader must have noticed that most of the conics appearing in our figures look like something that has been familiar ever since he first saw the full moon: they look like *circles*. This observation raises the important question: *Is the circle a conic*? We can answer *Yes* as soon as we have found a characteristic property of a conic that is also a property of the familiar circle. One way to do this is to give a Euclidean definition for the pole-and-polar relation with respect to a circle, and carry the discussion far enough to find that this relation satisfies the projective definition for a polarity. A quicker way is to give a Euclidean proof for the Braikenridge-MacLaurin construction (Figure 9.2B). More precisely, we select five points on a circle and prove that the unique conic that can be drawn through these points coincides with the circle. For convenience we shall take the five points A, P, B, Q, C

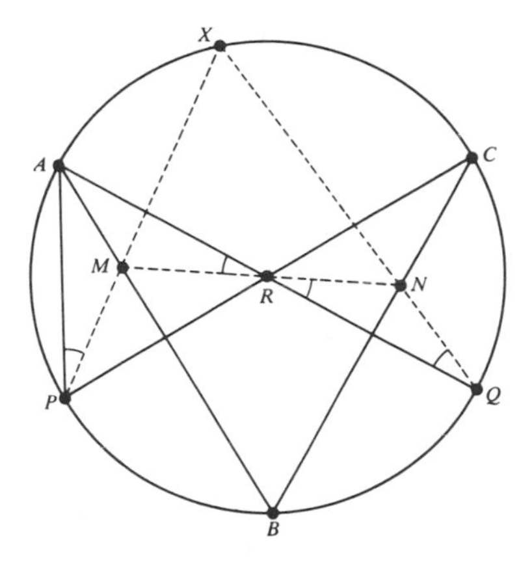

FIGURE 11.1A

(Figure 11.1A) to be five of the six vertices of a regular hexagon inscribed in the circle.

The following simple proof is due to S. L. Greitzer. Let AQ and CP meet in R (which is, of course, the center of the circle), let a variable diameter meet AB in M, BC in N, and let PM meet QN in X. Since AB, being a median of the equilateral triangle APR, is the perpendicular bisector of PR, and similarly BC is the perpendicular bisector of QR, we have

$$\angle XPA = \angle MPA = \angle ARM = \angle QRN = \angle NQR = \angle XQA.$$

By Euclid III.21 and 22, the locus of X is the circle APQ. By the Braikenridge-MacLaurin construction, the locus of X is the conic $APBQC$. Hence the conic coincides with the circle.

EXERCISES

1. What happens when the diameter is parallel to AB?
2. How does the proof have to be modified if A, P, B, Q, C are arbitrarily placed on the circle?

11.2 Affine Space

The time has come for us to investigate the connection between projective space, as determined by our axioms, and *affine* space, that is, the ordinary space of elementary solid geometry, in which two coplanar lines, or a line and a plane, or two planes, are said to be *parallel* if they do not meet. Instead of deliberately modifying the customary notions of space as we did in Section 1.1, let us now exploit the powerful theory of parallelism as Euclid did in his eleventh book. The idea of parallel lines leads immediately to that of a parallelogram, and thence to ratios of distances along parallel lines, or on one line (Reference 7, pp. 2, 115–128; Reference **8**, pp. 175, 202, 222).

I. M. Yaglom, in the third volume of his *Geometric Transformations* (Reference **23**, pp. 10, 21), shows two sketches of a house. The former depicts the living-room on a sunny day. The window panes cast shadows on the floor. The rectangles are distorted into parallelograms, but those that are equal remain equal. Affine geometry deals with properties that are maintained by such "parallel projection." His other sketch is a night scene outside the house. A lamp inside casts shadows of the same window panes on the lawn. The rectangles are distorted into quadrangles of various sizes, but their sides are still straight. Projective geometry deals with properties that are maintained by such "central projection."

One way of expressing the connection is to regard affine space as *part of* projective space, namely, projective space minus one plane. For this purpose, we specialize any one plane of the projective space and call it "the plane at infinity." (It is still, of course, a *projective* plane.) Two lines, or a line and a plane, or two planes, are then said to be parallel if they meet on this special plane. We soon see that parallelism, so defined, has all its familiar properties.

Another way is to begin with the affine space, regarding its properties as known, and seeking certain figures which behave like the projective points and lines. Such a representation of the projective *plane* is easy: we can use the lines and planes through a fixed point O in the affine space. For instance, to verify 2.22 we merely have to observe that any two distinct planes α and β, through O, have just one common line, namely the line $\alpha \cdot \beta$. For projective *space*, the appropriate figures are bundles and axial pencils, defined as follows.

A *bundle* is the set of all lines and planes through a point.

An *axial pencil* is the set of all planes through a line.

When there is any possibility of confusion, the other kind of pencil (the set of all lines that lie in a plane and pass through a point) is called a *flat* pencil. In the terminology of Section 1.5, an axial pencil, like a flat pencil, is a "one-dimensional form." But a bundle is a combination of two two-dimensional forms: the set of *lines* through a point (which is the space-dual of the set of *lines* in a plane) and the set of *planes* through the same point (which is the space-dual of the set of *points* in a plane). What makes these forms useful in the present connection is that their description is precisely the same in affine space as in projective space.

Clearly, a range is determined by any two of its points, and these may be any two distinct points; a flat pencil is determined by any two of its lines, and these may be any two lines that meet; an axial pencil is determined by any two of its planes, and these may be any two planes that meet; finally, a bundle (like a flat pencil) is determined by any two of its lines.

EXERCISE

A range, a flat pencil, and an axial pencil are three kinds of one-dimensional form. Which two are space-duals of each other? Which one is its own dual (that is, which one is "self-dual")?

11.3 How Two Coplanar Lines Determine a Flat Pencil and a Bundle

If (in ordinary space) we are given a point P, and two coplanar lines a and b whose point of intersection O is inconveniently far away, how can we construct the line through P of the pencil or bundle determined by a and b? If O were available we could simply draw OP, but can we still locate this line without using O? If P is not in the plane ab, we merely have to draw the planes Pa and Pb; these meet in a line p through P, which is the desired line of the bundle (see Figure 11.3A). In a single symbol, the member through P (outside the plane ab) is the line

$$p = Pa \cdot Pb.$$

On the other hand, if P is in the plane ab, we can use an auxiliary point Q outside the plane, locate the member q through Q (which is the line OQ), and consider (as in Figure 11.3B) the line of intersection of the planes ab and Pq. In other words, the line through P (of the bundle or pencil) is now

$$p = ab \cdot Pq \quad \text{where} \quad q = Qa \cdot Qb.$$

This construction owes its importance to the fact that it remains valid when a and b are parallel, so that O does not exist! The lines a, b, q, p, which originally passed through O, are now all parallel.

A practical model for this new situation is easily made by dividing a rectangular card into four unequal strips (of widths roughly proportional to

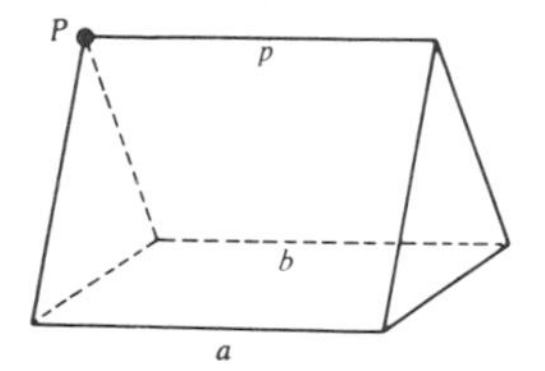

FIGURE 11.3A

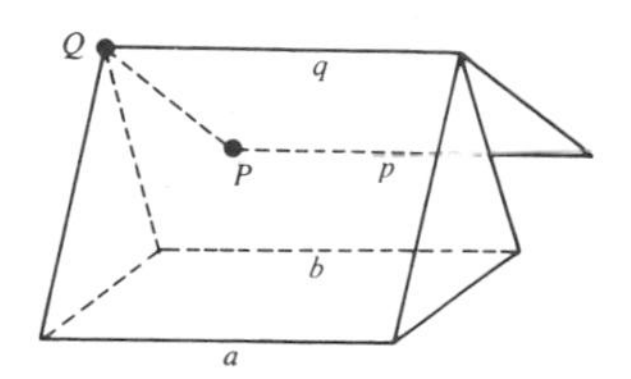

FIGURE 11.3B

4 : 6 : 7 : 5, as in Figure 11.3C) by three cuts made half-way through the thickness of the card. When suitably folded, this makes a solid version of Figure 11.3B.

Since we can derive p without inquiring whether a and b are parallel or not, we now feel justified in extending the meaning of the words *pencil* and *bundle* so as to allow the determining lines a and b to be *any two coplanar lines*. If a and b happen to be parallel, the bundle consists of all the lines and planes parallel to them, and the pencil consists of all the lines parallel to them in their own plane. Accordingly, we speak of a *bundle of parallels* and a (flat) *pencil of parallels*.

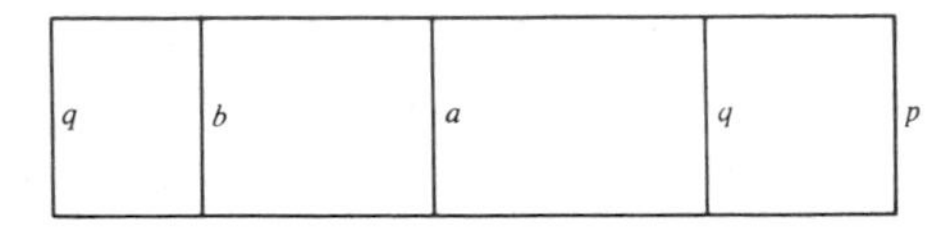

FIGURE 11.3C

It must be remembered that two planes may be parallel to a line without being parallel to each other. (For instance, in Figure 11.3B, the intersecting planes Qa and Qb are both parallel to the line p.) Thus a bundle of parallels contains a lot of lines, all parallel to one another, and a lot of planes, not all parallel to one another but each containing two (and therefore infinitely many) of the lines.

A familiar instance of a bundle of parallels is the set of all "vertical" lines and planes. If we take a cosmic standpoint and insist that two vertical lines are not strictly parallel but meet in the center of the earth, then we have an ordinary bundle instead of a bundle of parallels.

EXERCISE

Two lines parallel to the same line are parallel to each other. Does this remain true

(i) when the word "lines" is replaced by "planes,"
(ii) when the word "line" is replaced by "plane,"
(iii) when both substitutions are made simultaneously?

11.4 How Two Planes Determine an Axial Pencil

If we are given a point P, and two planes α and β whose line of intersection is far away, how can we construct the member through P of the axial pencil determined by α and β? This can be done by means of two applications of

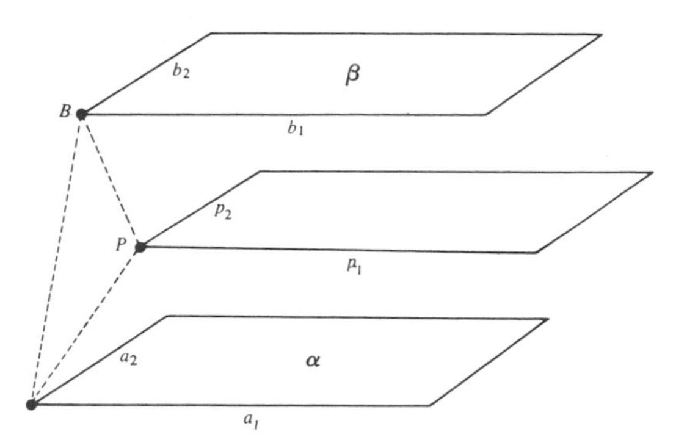

FIGURE 11.4A

the previous construction. We take any two intersecting lines a_1 and a_2 in α, and a point B in β (but not in either of the planes Pa_1, Pa_2), and draw the lines

$$b_1 = Ba_1 \cdot \beta, \qquad b_2 = Ba_2 \cdot \beta,$$
$$p_1 = Pa_1 \cdot Pb_1, \qquad p_2 = Pa_2 \cdot Pb_2,$$

as in Figure 11.4A. Then the desired plane through P is p_1p_2. For, if α and β meet in a line o, we may assume a_1 and a_2 to be chosen so as to meet o in two distinct points O_1 and O_2. Since $O_1 = o \cdot a_1$ lies in both the planes Ba_1 and β, it lies on their common line b_1. Since O_1 lies in both the planes Pa_1 and Pb_1, it lies on their common line p_1. Similarly O_2 lies on p_2. Therefore the join $o = O_1O_2$ lies in the plane p_1p_2.

If, on the other hand, α and β are parallel planes (conveyed by parallelograms in Figure 11.4A), the construction makes the lines b_1 and p_1 parallel to a_1, and the lines b_2 and p_2 parallel to a_2; therefore the plane p_1p_2 is parallel to α and β. Allowing P to take various positions, we thus obtain a *pencil of parallel planes*, consisting of all the planes parallel to a given plane.

A familiar instance is the set of all "horizontal"planes. If we insist that two horizontal planes are not strictly parallel but intersect in a line called the "horizon," then we have an ordinary axial pencil instead of a pencil of parallel planes.

EXERCISE

Can two lines be parallel to the same plane without being coplanar?

11.5 The Language of Pencils and Bundles

We have seen that a bundle can be derived from two of its lines, and an axial pencil from two of its planes, by constructions that remain valid when the two lines or planes are parallel.

Since an ordinary bundle consists of all the lines and planes through a point, and an ordinary axial pencil consists of all the planes through a line, any simple statement about points and lines can be "translated" into a corresponding statement about bundles and axial pencils. For instance, the statement

Any two distinct points lie on a line

becomes:

The common planes of any two distinct bundles form an axial pencil.

It is significant that the latter statement remains true when one of the bundles is replaced by a bundle of parallels, and again when *both* are replaced by bundles of parallels. In fact, the common planes of an ordinary bundle and a bundle of parallels form the ordinary axial pencil whose axis is the common line of the two bundles. (For instance, the bundle of lines and planes through any point O shares with the bundle of vertical lines and planes the vertical line through O and all the planes through this line.) Again, the common planes of two bundles of parallels form a pencil of parallel planes. (For instance, the bundle whose lines are horizontal in the north-south direction and the bundle whose lines are horizontal in the east-west direction have in common all the horizontal planes.)

EXERCISE

Translate the following statement into the language of pencils and bundles: *If two distinct lines have a common point they lie in a plane.*

11.6 The Plane at Infinity

These considerations serve to justify a convenient extension of space by the invention of an "ideal" plane whose points and lines represent the bundles of parallels and pencils of parallel planes, respectively. Remembering that an ordinary bundle consists of all the lines and planes through an ordinary point, we regard a bundle of parallels as consisting of all the lines and planes through an ideal point. Similarly, we regard a pencil of parallel planes as consisting of all the planes through an ideal line, and we say that an ideal point lies on an ideal line if the bundle contains the pencil. We can still assert that any two distinct points lie on a line. If one of the points is ordinary, so is the line; but if both are ideal, the line is ideal.

Since an ordinary bundle contains no pair of parallel planes, an ordinary point cannot lie on an ideal line; that is, all the "points" on an ideal line are ideal points. On the other hand, since a bundle of parallels contains ordinary axial pencils as well as pencils of parallel planes, an ideal point lies on some ordinary lines as well as on some ideal lines. Since any ordinary line belongs to just one bundle of parallels (consisting of all the lines and planes parallel to it), it contains just one ideal point, which we call its *point at infinity*. Thus we regard any two parallel lines as meeting in an ideal point: their common point at infinity. Since any plane belongs to just one pencil of parallel planes (consisting of all the planes parallel to it), it contains just one ideal line, which we call its *line at infinity*. Thus we regard any two parallel planes as meeting in an ideal line: their common line at infinity. In a given plane, each point on the line at infinity is the "center"of a pencil of parallel lines.

Since any two pencils of parallel planes belong to a bundle of parallels,

Any two ideal lines meet in an ideal point.

It follows that, if a and b are any two ideal lines, every other ideal line meets both a and b. This state of affairs resembles what happens in a plane. For, if a and b are two ordinary intersecting lines, every point in the plane ab lies on a line that meets both a and b. Accordingly, it is appropriate to regard the set of all ideal points and ideal lines as forming an ideal plane: *the plane at infinity*. This makes it possible to assert that any two intersecting (or parallel) lines determine a plane through both of them. If one of the lines is ordinary this is an ordinary plane; if both are ideal it is the plane at infinity.

Since each point (or line) at infinity is joined to an ordinary point O by an ordinary line (or plane), the points and lines of the projective *plane* may simply be regarded as a "new language" for the lines and planes (respectively) through O. In other words,

The projective plane is faithfully represented by a bundle.

Historically, points at infinity, lines at infinity, and the plane at infinity were first thought of by Kepler, Desargues, and Poncelet, respectively.

EXERCISE

Examine all the Axioms 2.11–2.18 for projective space, verifying that each is satisfied in our "extended" space, that is, in the ordinary affine space plus the plane at infinity with all its points and lines.

11.7 Euclidean Space

Although elementary solid geometry operates in affine space (Section 11.2), we must not imagine that affine geometry is merely another name for

Euclidean geometry! Affine geometry is the part of Euclidean geometry in which distances are compared only on the same line or on parallel lines.* Affine geometry becomes Euclidean geometry as soon as we have said what we mean by *perpendicular*. (For, right angles lead to circles and spheres, and thus enable us to compare distances.)

We have already referred to the set of all vertical lines as a familiar instance of a bundle of parallels, and to the set of all horizontal planes as a familiar instance of a pencil of parallel planes. More generally, every bundle of parallels in Euclidean space determines a unique axial pencil (of parallel planes), whose planes are perpendicular to the lines and planes of the bundle; and, conversely, every pencil of parallel planes determines a perpendicular bundle (of parallels). In the language of the plane at infinity, we thus have a special one-to-one correspondence between points at infinity and lines at infinity. As we have already remarked, the plane at infinity is a projective plane. Accordingly, we are not surprised to find that this correspondence between its points and lines is a polarity (called the *absolute polarity*). A line and a plane are perpendicular if the point at infinity on the line is the pole of the line at infinity in the plane. Two lines (or two planes) are perpendicular if their sections by the plane at infinity are conjugate points (or lines). Since no line or plane is perpendicular to itself, the polarity is elliptic. In fact, just as affine space can be derived from projective space by singling out a plane ("at infinity") and using it to define parallelism, so Euclidean space can be derived from affine space by singling out an elliptic polarity in the plane at infinity and using it to define perpendicularity.

EXERCISE

Let P and Q be two distinct points in Euclidean space. What is the locus of the point of intersection of a variable line through P and the perpendicular plane through Q?

---

* Melvin Hausner, "The Center of Mass and Affine Geometry," Am. Math. Monthly, **69** (1962), p. 730.

CHAPTER TWELVE

Coordinates

In any system of two-dimensional and homogeneous analytical geometry a point is a class of triads (x, y, z), those triads being classified together whose coordinates are proportional. . . . This is a very obvious observation, but it is of fundamental importance, since it marks the most essential difference between analytical geometries and "pure" geometries. . . . There are no axioms in any analytical geometry. An analytical geometry consists entirely of *definitions* and *theorems*.

G. H. Hardy
("What is geometry?" *Math. Gazette*, **12** (1925), p. 313)

12.1 The Idea of Analytic Geometry

Analytic geometry gives geometric names to certain sets of numbers in such a way that each geometric theorem is reduced to an algebraic theorem. Often the algebraic theorem is easy to prove, whereas the search for a "pure geometric" proof requires considerable ingenuity. For this reason many mathematicians prefer the algebraic method, thereby running the risk of producing a generation of students for whom a conic (for instance) is nothing more than a certain kind of quadratic equation. There is much to be said for an unbiassed attitude, in which each problem is solved by whichever method seems to work better. In earlier chapters we have concentrated on the pure or "synthetic" method; but now we restore the balance by showing how some of the same results can be obtained analytically.

For Euclidean geometry it is natural to use the classical "non-homogeneous" coordinates of Fermat, Descartes, and Newton, which may be illustrated by the description of a point in ordinary space (with reference to a chosen origin) as being at distances x_1 east, x_2 north, and x_3 up. For projective geometry it is more convenient to use the homogeneous coordinates of Möbius, Grassmann, and Plücker, which may be illustrated by the description of a point in a plane (with reference to a triangle $A_1A_2A_3$) as being at the center of gravity of masses x_1 at A_1, x_2 at A_2, and x_3 at A_3. However, such illustrations are not at all essential; the important idea is to take an ordered set of numbers (x_1, x_2, x_3) and call it a point.

The "numbers" that we use may for simplicity be thought of as *real* numbers, but actually they can be the elements of any commutative field (Reference **15**, p. 21) in which $1 + 1 \neq 0$; in particular, they can form a finite field, and this throws light on the subject of Chapter 10.

In order to be able to interpret lines as well as points, we consider two types of ordered triads of numbers, say (x_1, x_2, x_3) and $[X_1, X_2, X_3]$. We agree to exclude the "trivial" triads (0, 0, 0), [0, 0, 0], and to regard two triads of the same type as being equivalent (that is, geometrically indistinguishable) if they are proportional; thus (x_1, x_2, x_3) is equivalent to $(\lambda x_1, \lambda x_2, \lambda x_3)$, and $[X_1, X_2, X_3]$ is equivalent to $[\lambda X_1, \lambda X_2, \lambda X_3]$, for any $\lambda \neq 0$. With two triads of opposite types we associate a single number, their "inner product"

(12.11) $$\{xX\} = \{Xx\} = X_1x_1 + X_2x_2 + X_3x_3,$$

which may be zero.

EXERCISE

In Section 11.2, we considered the possibility of representing the points and lines of the projective plane by the lines and planes through a fixed point O in affine space. How is this representation related to homogeneous and non-homogeneous coordinates?

12.2 Definitions

Setting aside all ideas of measurement, let us now build up the analytic geometry of the projective plane in the manner proposed by Hardy, that is, as consisting entirely of definitions and theorems, beginning with the definitions of point, line, and incidence.

A *point* is the set of all triads equivalent to a given triad (x_1, x_2, x_3). In other words, a point is an ordered set of three numbers (x_1, x_2, x_3), not all zero, with the understanding that $(\lambda x_1, \lambda x_2, \lambda x_3)$ is the same point for any nonzero λ. For instance, (2, 3, 6) is a point, and $(-\frac{1}{3}, -\frac{1}{2}, -1)$ is another way of writing the same point.

A *line* is the set of all triads equivalent to a given triad $[X_1, X_2, X_3]$. In other words, a line is defined in the same manner as a point, but with square brackets instead of ordinary parentheses, and with capital letters to represent the three numbers. Thus [3, 2, −2] is a line, and $[-1, -\frac{2}{3}, \frac{2}{3}]$ is the same line. We shall find that the point (x_1, x_2, x_3) and the line $[x_1, x_2, x_3]$ (with the same x's) are related by a polarity, but the apparently special role of this polarity is a notational accident.

The point (x) and line $[X]$, meaning (x_1, x_2, x_3) and $[X_1, X_2, X_3]$, are said to be *incident* (the point lying on the line and the line passing through the point) if and only if

$$\{xX\} = 0$$

in the notation of (12.11). For instance, (2, 3, 6) lies on [3, 2, −2]. It follows that any discussion can be dualized by interchanging small and capital letters, round and square brackets.

The three numbers x_i are called the *coordinates* (or "homogeneous coordinates," or "projective coordinates") of the point (x). The three numbers X_i are called the *coordinates* (or "line coordinates," or "envelope coordinates," or "tangential coordinates") of the line $[X]$.

If (x) is a variable point on a fixed line $[X]$, we call $\{Xx\} = 0$ the *equation* of the line $[X]$, because it is a characteristic property of points on the line. For instance, the line [3, 2, −2] has the equation

$$3x_1 + 2x_2 - 2x_3 = 0, \quad \text{or} \quad 3x_1 + 2x_2 = 2x_3.$$

Dually, if $[X]$ is a variable line through a fixed point (x), we call $\{xX\} = 0$ (which is the same as $\{Xx\} = 0$) the *equation* of the point (x), because it is a characteristic property of lines through the point. For instance, the point (2, 3, 6) has the equation

$$2X_1 + 3X_2 + 6X_3 = 0,$$

and the point (1, 0, 0) has the equation $X_1 = 0$. Thus the coordinates of a line or point are the coefficients in its equation (with zero for any missing term).

The three points (1, 0, 0), (0, 1, 0), (0, 0, 1), or $X_i = 0$ $(i = 1, 2, 3)$, and the three lines [1, 0, 0], [0, 1, 0], [0, 0, 1], or $x_i = 0$, are evidently the vertices and sides of a triangle. We call this the *triangle of reference* (see Figure 12.2A). The point (1, 1, 1) and line [1, 1, 1] are called the *unit point* and *unit line*. We shall find, in Section 12.4, that there is nothing geometrically special about this triangle and point and line, except that the point does not lie on a side, the line does not pass through a vertex, and the point is the trilinear pole of the line (see Section 3.4).

By eliminating X_1, X_2, X_3 from the equations

$$\{xX\} = 0, \qquad \{yX\} = 0, \qquad \{zX\} = 0$$

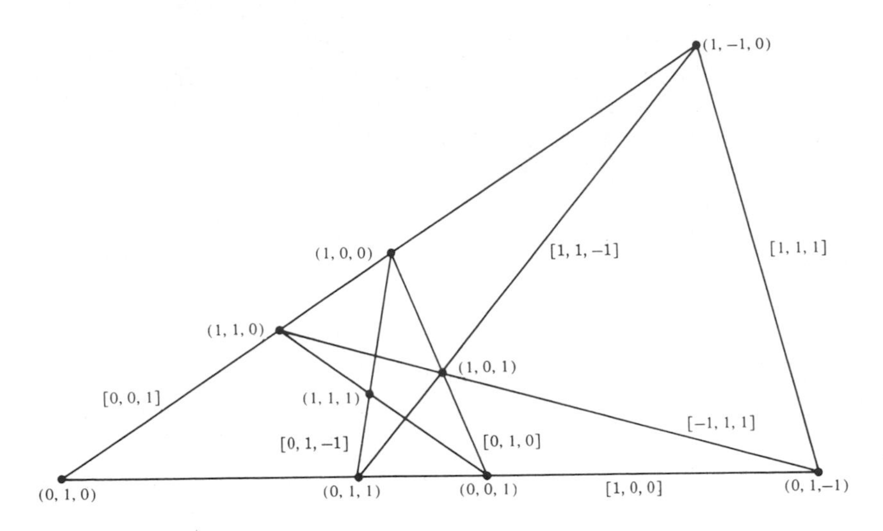

FIGURE 12.2A

of three given points (x), (y), (z), we find the necessary and sufficient condition

(12.21)
$$\begin{vmatrix} x_1 & x_2 & x_3 \\ y_1 & y_2 & y_3 \\ z_1 & z_2 & z_3 \end{vmatrix} = 0$$

for the three points to be collinear. This condition is equivalent to the existence of numbers λ, μ, ν, not all zero, such that

$$\lambda x_i + \mu y_i + \nu z_i = 0 \qquad (i = 1, 2, 3).$$

If (y) and (z) are distinct points, $\lambda \neq 0$. Hence the general point collinear with (y) and (z) is $(\mu y_1 + \nu z_1, \mu y_2 + \nu z_2, \mu y_3 + \nu z_3)$ or, briefly,

$$(\mu y + \nu z)$$

where μ and ν are not both zero.

When $\nu = 0$, this is the point (y) itself. For any other position, since (νz) is the same point as (z), we can allow the coordinates of (z) to absorb the ν, and the point collinear with (y) and (z) is simply

$$(\mu y + z).$$

If we are concerned with only one such point, we may allow the μ to be absorbed too; thus three distinct collinear points may be expressed as (y), (z),

$(y + z)$. However, this last simplification cannot be effected simultaneously on two lines if thereby one point would have to absorb two different parameters. The symbol $(\mu y + z)$ can be made to include *every* point on the line $(y)(z)$ if we accept the convention that the point (y) is $(\mu y + z)$ with $\mu = \infty$. Dually, the condition for three lines $[X]$, $[Y]$, $[Z]$ to be concurrent is

$$\begin{vmatrix} X_1 & X_2 & X_3 \\ Y_1 & Y_2 & Y_3 \\ Z_1 & Z_2 & Z_3 \end{vmatrix} = 0; \tag{12.22}$$

the general line concurrent with $[Y]$ and $[Z]$ is $[\mu Y + \nu Z]$; and any particular line concurrent with $[Y]$ and $[Z]$, but distinct from them, may be expressed as $[Y + Z]$.

EXERCISES

1. Where does the unit line $[1, 1, 1]$ meet the sides of the triangle of reference?

2. Copy Figure 12.2A (without the coordinate symbols) and add the lines $[0, 1, 1]$, $[-1, 0, 1]$, $[1, 1, 0]$. Are these lines concurrent?

3. What line joins the points $(1, 1, 1)$ and $(1, -1, 0)$? Where does it meet $[1, 0, 0]$?

4. The line joining $(1, 0, 0)$ to (x_1, x_2, x_3) is $[0, x_3, -x_2]$. Where does it meet $[1, 0, 0]$?

5. If the triangle of reference is the diagonal triangle of a quadrangle having $(1, 1, 1)$ for one vertex, where are the other three vertices? (See Exercise 1 of Section 3.3.)

6. The lines $[X]$ and $(y)(z)$ meet in the point
$$(\{Xz\}y - \{Xy\}z).$$
[*Hint:* What is the condition for $(\mu y + z)$ to lie on $[X]$?]

7. Obtain coordinates for the various points and lines in Figure 3.4A, beginning with
$$A = (1, 0, 0), \qquad B = (0, 1, 0), \qquad C = (0, 0, 1),$$
$$S = (x_1, x_2, x_3).$$
Deduce the condition $x_1X_1 = x_2X_2 = x_3X_3$ for the point (x) and line $[X]$ to be trilinear pole and polar with respect to the triangle of reference.

8. Where is the harmonic conjugate of $(x_1, x_2, 0)$ with respect to $(1, 0, 0)$ and $(0, 1, 0)$?

12.3 Verifying the Axioms for the Projective Plane

To show that this analytic geometry provides a model for the synthetic geometry developed in earlier chapters, we must verify that Axioms 2.13, 3.11, 3.12, 2.17, 2.18 (see page 95) are all satisfied.

The first two can be verified as follows. Two points (y) and (z) are joined by the line (12.21) or

$$\left[\begin{vmatrix} y_2 & y_3 \\ z_2 & z_3 \end{vmatrix}, \begin{vmatrix} y_3 & y_1 \\ z_3 & z_1 \end{vmatrix}, \begin{vmatrix} y_1 & y_2 \\ z_1 & z_2 \end{vmatrix}\right].$$

Two lines $[Y]$ and $[Z]$ meet in the point (12.22) or

$$\left(\begin{vmatrix} Y_2 & Y_3 \\ Z_2 & Z_3 \end{vmatrix}, \begin{vmatrix} Y_3 & Y_1 \\ Z_3 & Z_1 \end{vmatrix}, \begin{vmatrix} Y_1 & Y_2 \\ Z_1 & Z_2 \end{vmatrix}\right).$$

To verify Axioms 3.12 and 2.17, we consider a quadrangle $PQRS$ whose first three vertices (p), (q), (r) satisfy

(12.31)
$$\begin{vmatrix} p_1 & p_2 & p_3 \\ q_1 & q_2 & q_3 \\ r_1 & r_2 & r_3 \end{vmatrix} \neq 0.$$

Since the side PS joins (p) to the diagonal point $A = QR \cdot PS$, we may take A (on QR, but distinct from Q and R) to be $(q + r)$, and S (on PA, but distinct from P and A) to be $(p + q + r)$, meaning

$$(p_1 + q_1 + r_1,\ p_2 + q_2 + r_2,\ p_3 + q_3 + r_3).$$

Then B, on both RP and QS, must be $(r + p)$, and C, on both PQ and RS, must be $(p + q)$, The three diagonal points A, B, C are noncollinear since

(12.32)
$$\begin{vmatrix} q_1 + r_1 & q_2 + r_2 & q_3 + r_3 \\ r_1 + p_1 & r_2 + p_2 & r_3 + p_3 \\ p_1 + q_1 & p_2 + q_2 & p_3 + q_3 \end{vmatrix} = 2\begin{vmatrix} p_1 & p_2 & p_3 \\ q_1 & q_2 & q_3 \\ r_1 & r_2 & r_3 \end{vmatrix} \neq 0.$$

Similar ideas provide a simple proof for Desargues's theorem (cf. page 38). Let the first triangle PQR and the center of perspective O be $(p)(q)(r)$ and (u). There is no loss of generality in taking the second triangle $P'Q'R'$ to be

$$(p + u)(q + u)(r + u).$$

The point $D = QR \cdot Q'R'$, being collinear with (q) and (r) and also with $(q + u)$ and $(r + u)$, can only be $(q - r)$. Similarly, E is $(r - p)$, and F is $(p - q)$. These points D, E, F are collinear since

$$\begin{vmatrix} q_1 - r_1 & q_2 - r_2 & q_3 - r_3 \\ r_1 - p_1 & r_2 - p_2 & r_3 - p_3 \\ p_1 - q_1 & p_2 - q_2 & p_3 - q_3 \end{vmatrix} = 0$$

or, more simply, since

$$(q_i - r_i) + (r_i - p_i) + (p_i - q_i) = 0 \qquad (i = 1, 2, 3).$$

When a range of points P arises as a section of a pencil of lines p, the "elementary correspondence" $P \barwedge p$ may be described as relating three positions of P, say

$$(y), \qquad (z), \qquad (y + z)$$

to three positions of p, say

$$[Y], \qquad [Z], \qquad [Y + Z].$$

From the information that P and p are incident in these three cases, can we deduce that, when P is $(\mu y + z)$, p is $[\mu Y + Z]$ with the same μ? Yes! Since

$$\{(Y + Z)(y + z)\} = \{Yy\} + \{Yz\} + \{Zy\} + \{Zz\},$$

the three given incidences imply

$$\{Yy\} = 0, \qquad \{Zz\} = 0, \qquad \{Yz\} + \{Zy\} = 0,$$

whence

$$\{(\mu Y + Z)(\mu y + z)\} = \mu^2\{Yy\} + \mu(\{Yz\} + \{Zy\}) + \{Zz\} = 0,$$

showing that the line $[\mu Y + Z]$ is indeed incident with the point $(\mu y + z)$.

Repeated application of this result shows that the relation

$$(y)(z)(y + z)(\mu y + z) \barwedge [Y][Z][Y + Z][\mu Y + Z]$$

holds not only for an elementary correspondence but for any projectivity from a range to a pencil; and of course we have also

$$(y)(z)(y + z)(\mu y + z) \barwedge (y')(z')(y' + z')(\mu y' + z'),$$

$$[Y][Z][Y + Z][\mu Y + Z] \barwedge [Y'][Z'][Y' + Z'][\mu Y' + Z'].$$

This is the algebraic version of the Fundamental Theorem 4.12, from which Axiom 2.18 can be deduced as a special case.

It is interesting to compare this verification of the axioms with what we did in Section 10.3. An important difference is that, whereas $PG(2, 5)$ is a single (categorical) geometry, the analytic geometry that we are discussing now has the same degree of freedom as the synthetic geometry itself. More advanced treatises give synthetic proofs that the points on a line can be "added" and "multiplied" so as to constitute the elements of a field. (Chapters 7 and 11 of Reference 7 remain valid when all references to order and continuity have been omitted.) In other words, projective geometry becomes categorical as soon as the field of coordinates has been specified.

EXERCISES

1. The harmonic conjugate of $(y + z)$ with respect to (y) and (z) is $(y - z)$.
2. Where is the harmonic conjugate of $(\mu y + z)$ with respect to (y) and (z)? Dualize this result.
3. The hyperbolic involution with invariant points (y) and (z) is
$$(\mu y + z) \;\overline{\wedge}\; (-\mu y + z).$$
4. Any projectivity on the line $(y)(z)$ is expressible in the form
$$(\mu y + z) \;\overline{\wedge}\; (\mu' y + z),$$
where $\mu' = (\alpha\mu + \beta)/(\gamma\mu + \delta)$, $\alpha\delta \neq \beta\gamma$.
5. Give an equation (or equations) for the general projectivity on the line $[0, 0, 1]$.
6. Under what circumstances will the projectivity described in Exercise 4 or 5 be:
(i) an involution, (ii) parabolic?
7. When four distinct collinear points are expressed in the form
$$(y), \qquad (z), \qquad (y + z), \qquad (\mu y + z),$$

the number μ is called the *cross ratio* of the four points. There is an analogous definition for the cross ratio of four concurrent lines. Two such tetrads are projectively related if and only if they have the same cross ratio.

8. For two points (y), (z) and two lines $[Y]$, $[Z]$, the expression
$$\frac{\{yY\}\{zZ\}}{\{yZ\}\{zY\}}$$

is equal to the cross ratio of the two given points with the points in which their join meets the two lines.

9. The cross ratio of $(a, 1, 0)$, $(b, 1, 0)$, $(c, 1, 0)$, $(d, 1, 0)$ is

$$\frac{(a - c)(b - d)}{(a - d)(b - c)}.$$

12.4 Projective Collineations

We have seen that the condition (12.21) makes the points (x), (y), (z) collinear; conversely (12.31) makes (p), (q), (r) noncollinear, so that they form a triangle. This triangle enables us to describe the position of any point by means of *barycentric* coordinates λ, μ, ν, which are the coefficients in the expression $(\lambda p + \mu q + \nu r)$. This is an obvious generalization of the expression $(p + q + r)$ used in the proof of Axiom 2.17 in Section 12.3. Points on a side of the triangle can be included by allowing $\lambda\mu\nu = 0$, and when $\mu = \nu = 0$ we have the point (p) itself.

When $(p)(q)(r)$ and $(p + q + r)$ are the triangle of reference and unit point, $(\lambda p + \mu q + \nu r)$ is (λ, μ, ν), and the barycentric coordinates are the same as the ordinary coordinates.

The correspondence $(x) \rightarrow (x')$, where

$$\text{(12.41)} \qquad \begin{cases} x'_1 = p_1x_1 + q_1x_2 + r_1x_3, \\ x'_2 = p_2x_1 + q_2x_2 + r_2x_3, \\ x'_3 = p_3x_1 + q_3x_2 + r_3x_3, \end{cases}$$

transforms (λ, μ, ν) into the point $(\lambda p + \mu q + \nu r)$ which has these same barycentric coordinates referred to $(p)(q)(r)$ instead of $(1, 0, 0)(0, 1, 0)(0, 0, 1)$.

Since we are assuming (12.31), so that the equations (12.41) can be solved for the x's in terms of the x''s, this is a point-to-point correspondence as described at the beginning of Section 6.1. Since the equation $\{X'x'\} = 0$ [see (12.11)] is equivalent to

$$\text{(12.42)} \qquad \{X'p\}x_1 + \{X'q\}x_2 + \{X'r\}x_3 = 0,$$

it is a collineation. Since it transforms $(0, \mu, \nu)$ into $(\mu q + \nu r)$, the collineation is projective (see 6.11). Since it transforms the quadrangle

$$(1, 0, 0)(0, 1, 0)(0, 0, 1)(1, 1, 1)$$

into $(p)(q)(r)(p + q + r)$, which may be identified with *any* given quadrangle by a suitable choice of the p's and q's and r's, it is the *general* projective collineation (see 6.13).

This collineation, which shifts the points (x) to new positions (x'), is the active or *alibi* aspect of the linear homogeneous transformation (12.41). There is also a passive or *alias* aspect: a coordinate transformation that gives a new name to each point. In fact, we may regard $(p)(q)(r)$ as a new triangle of reference, with respect to which the point that we have been calling (x') has coordinates

$$(x_1, x_2, x_3),$$

whereas its coordinates with respect to the original triangle are, of course, (x'_1, x'_2, x'_3).

One practical consequence of the "alias" aspect is that, when seeking an analytic proof for a theorem concerning a triangle and a point of general position, we are justified in using the triangle of reference and unit point. Similarly, for a theorem concerning a quadrangle, it is often convenient to take the vertices to be $(1, \pm 1, \pm 1)$, so that the six sides have equations $x_i \pm x_j = 0$ $(i < j)$ and the diagonal triangle is the triangle of reference. Dually, a given quadrilateral may be taken to have sides $[1, \pm 1, \pm 1]$ and vertices $X_i \pm X_j = 0$.

Of course, a collineation is not only a point-to-point transformation but also a line-to-line transformation. The latter aspect of the collineation (12.41) is, by (12.42),

$$X_1 = \{pX'\}, \qquad X_2 = \{qX'\}, \qquad X_3 = \{rX'\}.$$

A more systematic notation for the same two sets of equations is

(12.43)
$$\rho x'_i = c_{i1}x_1 + c_{i2}x_2 + c_{i3}x_3 = \Sigma\, c_{ij}x_j \qquad (i = 1, 2, 3),$$
$$\sigma X_j = c_{1j}X'_1 + c_{2j}X'_2 + c_{3j}X'_3 = \Sigma\, c_{ij}X'_i \qquad (j = 1, 2, 3),$$

where $\rho\sigma \neq 0$ and, by (12.31),

(12.44)
$$\begin{vmatrix} c_{11} & c_{12} & c_{13} \\ c_{21} & c_{22} & c_{23} \\ c_{31} & c_{32} & c_{33} \end{vmatrix} \neq 0$$

(Reference **19**, p. 187; Reference **17**, pp. 68, 85). The preservation of incidence is verified as follows:

$$\rho\{X'x'\} = \rho\Sigma\, X'_i x'_i = \Sigma\Sigma\, c_{ij}X'_i x_j = \sigma\Sigma\, X_j x_j = \sigma\{Xx\}.$$

Since our coordinates are homogeneous, there are many occasions when we can omit the ρ and σ in (12.43), that is, set $\rho = \sigma = 1$. However, it is important to retain them when we are looking for invariant points or invariant lines. The invariant points are naturally given by $x'_i = x_i$ or

$$\rho x_i = \Sigma\, c_{ij} x_j \qquad (i = 1, 2, 3).$$

Eliminating the x's from these three equations, we obtain

$$\begin{vmatrix} c_{11} - \rho & c_{12} & c_{13} \\ c_{21} & c_{22} - \rho & c_{23} \\ c_{31} & c_{32} & c_{33} - \rho \end{vmatrix} = 0.$$

Any root ρ of this *characteristic equation* makes the three equations for the x's consistent, and then we can solve any two of them to obtain the coordinates of an invariant point.

For instance, the collineation

$$\textbf{(12.45)} \qquad \rho x'_1 = x_1, \quad \rho x'_2 = x_2, \quad \rho x'_3 = \mu^{-1} x_3 \qquad (\mu \neq 1)$$

has the characteristic equation $(\rho - 1)^2(\rho - \mu^{-1}) = 0$. The double root $\rho = 1$ yields the range of invariant points $(x_1, x_2, 0)$, and the remaining root $\rho = \mu^{-1}$ yields the isolated invariant point $(0, 0, 1)$. By 6.24 and 6.27, this collineation is a homology with center $(0, 0, 1)$ and axis $[0, 0, 1]$.

Again, the collineation

$$\textbf{(12.46)} \qquad \rho x'_1 = x_1 + a_1 x_3, \quad \rho x'_2 = x_2 \quad a_2 x_3, \quad \rho x'_3 = x_3$$

has the characteristic equation $(\rho - 1)^3 = 0$. If a_1 and a_2 are not both zero, the triple root $\rho = 1$ yields the range of invariant points $(x_1, x_2, 0)$; there are no others. By 6.24 and 6.26, this collineation is an elation with axis $[0, 0, 1]$. Since the equation $a_2 x'_1 - a_1 x'_2 = 0$ implies $a_2 x_1 - a_1 x_2 = 0$, there is an invariant line (other than $[0, 0, 1]$) through the point $(a_1, a_2, 0)$. Hence this point is the center of the elation.

Comparing the two parts of (12.43), we see that the expression for the homology (12.45) as a line-to-line transformation is

$$\sigma X_1 = X'_1, \quad \sigma X_2 = X'_2, \quad \sigma X_3 = \mu^{-1} X'_3$$

or, taking $\sigma = 1$ for convenience and solving,

$$\textbf{(12.47)} \qquad X'_1 = X_1, \quad X'_2 = X_2, \quad X'_3 = \mu X_3.$$

Similarly, the elation (12.46), *qua* line-to-line transformation, is

$$\textbf{(12.48)} \qquad X'_1 = X_1, \quad X'_2 = X_2, \quad X'_3 = X_3 - a_1 X_1 - a_2 X_2.$$

EXERCISES

1. Find the projective collineations that transform

$$(1, 0, 0)(0, 1, 0)(0, 0, 1)(1, 1, 1)$$

into the following quadrangles:

(i) $(1, 0, 0)(0, 1, 0)(0, 0, 1)(p, q, r)$,
(ii) $(-1, 1, 1)(1, -1, 1)(1, 1, -1)(1, 1, 1)$,
(iii) $(0, 1, 0)(0, 0, 1)(1, 0, 0)(1, 1, 1)$,
(iv) $(0, 1, 0)(0, 0, 1)(1, 1, 1)(1, 0, 0)$.

2. Give equations for the inverse of the collineation (12.43).
3. Prove Desargues's theorem as applied to the triangle of reference and $(p, 1, 1)(1, q, 1)(1, 1, r)$.
4. Prove Pappus's theorem. [*Hint:* Use Exercise 2 of Section 4.4, taking $A_1A_2A_3$ as triangle of reference and $B_1 = (p, 1, 1)$, $B_2 = (1, 1, r)$.]

12.5 Polarities

Since the product of two correlations (for example, a polarity and another projective correlation) is a collineation, any given projective correlation can be exhibited as the product of an arbitrary polarity and a suitable projective collineation. The most convenient polarity for this purpose is the one that transforms each point (or line) into the line (or point) that has the *same* coordinates. (This correlation is obviously projective, and of period 2.) Combining the general projective collineation (12.43) with the polarity that interchanges X'_i and x'_i, we obtain the general projective correlation in the form

(12.51)

$$\rho X'_i = c_{i1}x_1 + c_{i2}x_2 + c_{i3}x_3 = \Sigma\, c_{ij}x_j \qquad (i = 1, 2, 3),$$

$$\sigma X_j = c_{1j}x'_1 + c_{2j}x'_2 + c_{3j}x'_3 = \Sigma\, c_{ij}x'_i \qquad (j = 1, 2, 3),$$

where again the coefficients satisfy (12.44). Incidences are dualized in the proper manner for a correlation, since

$$\rho\{X'x'\} = \rho\Sigma\, X'_ix'_i = \Sigma\Sigma\, c_{ij}x'_ix_j = \sigma\Sigma\, X_jx_j = \sigma\{Xx\}.$$

The projective correlation (12.51) is a *polarity* if it is equivalent to the inverse correlation $\sigma X'_j = \Sigma\, c_{ij}x_i$ or (interchanging i and j)

$$\sigma X'_i = \Sigma\, c_{ji}x_j.$$

This means that

$$c_{ji} = \frac{\sigma}{\rho}\, c_{ij}$$

with the same σ/ρ for all i and j, so that, since the c_{ij} are not all zero,

$$c_{ij} = \frac{\sigma}{\rho} c_{ji} = \left(\frac{\sigma}{\rho}\right)^2 c_{ij}, \quad \left(\frac{\sigma}{\rho}\right)^2 = 1, \quad \frac{\sigma}{\rho} = \pm 1.$$

The lower sign is inadmissible, as that would make $c_{ji} = -c_{ij}$ and

$$\begin{vmatrix} c_{11} & c_{12} & c_{13} \\ c_{21} & c_{22} & c_{23} \\ c_{31} & c_{32} & c_{33} \end{vmatrix} = \begin{vmatrix} 0 & c_{12} & -c_{31} \\ -c_{12} & 0 & c_{23} \\ c_{31} & -c_{23} & 0 \end{vmatrix} = 0.$$

Hence $\sigma = \rho$ and $c_{ji} = c_{ij}$. In other words, a projective correlation is a polarity if and only if the matrix of coefficients c_{ij} is *symmetric*.

The nature of a polarity is such that no confusion can be caused by omitting the prime (as in Section 7.1) and writing simply

(12.52) $$X_i = \Sigma\, c_{ij} x_j \qquad (i = 1, 2, 3),$$

where $c_{ij} = c_{ji}$ and $\det(c_{ij}) = \Delta \neq 0$. These equations give us the polar $[X]$ of a given point (x). Solving them, we obtain the pole (x) of a given line $[X]$ in the form

$$\Delta x_i = \Sigma\, C_{ij} X_j \qquad (i = 1, 2, 3),$$

where C_{ij} is the cofactor of c_{ij} in the determinant Δ.

Two points (x) and (y) are conjugate if (x) lies on the polar $[Y]$ of (y). Since $Y_i = \Sigma\, c_{ij} y_j$, the condition $\{Yx\} = 0$ or $\Sigma\, Y_i x_i = 0$ becomes

(12.53) $$\Sigma\Sigma\, c_{ij} x_i y_j = 0,$$

which we shall sometimes write in the abbreviated form

$$(xy) = 0.$$

Letting (x) vary, we see that this is *the equation of the polar of* (y). Dually, the condition for lines $[X]$ and $[Y]$ to be conjugate, or the equation of the pole of $[Y]$, is

(12.54) $$[XY] = 0,$$

where $[XY] = \Sigma\Sigma\, C_{ij} X_i Y_j$.

As a particular case of $(xy) = 0$, the condition for $(0, 1, 0)$ and $(0, 0, 1)$ to be conjugate is $c_{23} = 0$. Thus the triangle of reference is self-polar if and only if

$$c_{23} = c_{31} = c_{12} = 0.$$

By choosing any self-polar triangle as triangle of reference, we reduce a given polarity to its *canonical form*

$$X_i = c_{ii} x_i \qquad (c_{11} c_{22} c_{33} \neq 0)$$

or, more conveniently,

(12.55) $$X_1 = ax_1, \quad X_2 = bx_2, \quad X_3 = cx_3 \qquad (abc \neq 0).$$

This is the polarity $(ABC)(Pp)$, where ABC is the triangle of reference, P is $(1, 1, 1)$, and p is $[a, b, c]$.

EXERCISES

1. Prove 7.21, using the triangle of reference.
2. Prove 7.13, using the polarity (12.52) and the line $[0, 0, 1]$.
3. Prove 7.61 (Hesse's theorem), using the quadrilateral $[1, \pm 1, \pm 1]$, whose pairs of opposite vertices are $(\pm 1, 1, 0)$, $(0, \pm 1, 1)$, $(\pm 1, 0, 1)$.
4. Prove 7.31 (Chasles's theorem), using the triangle of reference and the general polarity.
5. Triangles $(0,1,1)\ (1,0,1)\ (1,1,0)$ and $(-1, 1, 1)\ (1, -1, 1)\ (1, 1, -1)$ are perspective from $(1, 1, 1)$ and $[1, 1, 1]$. For what polarity are they polar triangles?
6. In the terminology of Exercise 6 at the end of Chapter 10, a point (y) is *accessible* from (z), with respect to the polarity (12.52), if and only if *there is a number* $\mu \neq 0$ *such that*
$$(yy) = \mu^2(zz):$$
either (yy) and (zz) are both zero or their product is a nonzero *square*. (In real geometry this means that (yy) and (zz) have the same sign. But in *complex* geometry, since every complex number is a square, all non-self-conjugate points are mutually accessible.)

12.6 Conics

The condition for a point (x) to be self-conjugate for the polarity (12.52) is $(xx) = 0$, or
$$c_{11}x_1^2 + c_{22}x_2^2 + c_{33}x_3^2 + 2c_{23}x_2x_3 + 2c_{31}x_3x_1 + 2c_{12}x_1x_2 = 0.$$
Dually, the condition for a line $[X]$ to be self-conjugate is $[XX] = 0$, or
$$C_{11}X_1^2 + C_{22}X_2^2 + C_{33}X_3^2 + 2C_{23}X_2X_3 + 2C_{31}X_3X_1 + 2C_{12}X_1X_2 = 0.$$
Hence every conic (locus or envelope) has such an equation. In particular, using (12.55) instead of (12.52), every conic for which the triangle of reference is self-polar has an equation of the form
$$ax_1^2 + bx_2^2 + cx_3^2 = 0 \quad \text{or} \quad a^{-1}X_1^2 + b^{-1}X_2^2 + c^{-1}X_3^2 = 0.$$

We can now clarify the statement (in Section 8.1) that in some geometries every polarity is hyperbolic, whereas other geometries admit elliptic polarities too. The polarity (12.52) is hyperbolic or elliptic according as the equation $(xx) = 0$ does or does not have a solution (other than $x_1 = x_2 = x_3 = 0$). The distinction depends on the coordinate field. If this is the field of *complex* numbers, every such equation can be solved; for example, the equation

$$x_1^2 + x_2^2 + x_3^2 = 0$$

is satisfied by $(1, \sqrt{-1}, 0)$. Over such a field, every polarity is hyperbolic. In the case of the field of *real* numbers, on the other hand, the quadratic form (xx) may be "definite," in which case the polarity (for instance, $X_i = x_i$) is elliptic. (The only solution of the above equation in real numbers is 0, 0, 0.)

Some particular equations represent conics regardless of the field. For example, the equation

(12.61) $$x_1^2 + x_2^2 - x_3^2 = 0,$$

being satisfied by (1, 0, 1), cannot fail to represent a conic.

Since the condition for the conic $(xx) = 0$ to pass through (1, 0, 0) is $c_{11} = 0$, the most general conic circumscribing the triangle of reference is

$$c_{23}x_2x_3 + c_{31}x_3x_1 + c_{12}x_1x_2 = 0 \qquad (c_{23}c_{31}c_{12} \neq 0).$$

The coordinate transformation

$$x_1 \rightarrow c_{23}x_1, \qquad x_2 \rightarrow c_{31}x_2, \qquad x_3 \rightarrow c_{12}x_3$$

converts this into

(12.62) $$x_2x_3 + x_3x_1 + x_1x_2 = 0.$$

Thus, in any problem concerning a triangle and a circumscribed conic, the conic can be expressed in this simple form. Working out the cofactors in the determinant, we obtain the envelope equation

$$X_1^2 + X_2^2 + X_3^2 - 2X_2X_3 - 2X_3X_1 - 2X_1X_2 = 0$$

or

$$X_1^{\frac{1}{2}} \pm X_2^{\frac{1}{2}} \pm X_3^{\frac{1}{2}} = 0.$$

Dually, a conic inscribed in the triangle of reference is

$$X_2X_3 + X_3X_1 + X_1X_2 = 0$$

or

(12.63) $$x_1^{\frac{1}{2}} \pm x_2^{\frac{1}{2}} \pm x_3^{\frac{1}{2}} = 0.$$

EXERCISES

1. Prove 8.11, using the conic $x_2x_3 + x_3x_1 + x_1x_2 = 0$.

2. Prove 8.41, taking the points P, Q, R and tangents p, q to be $(1, 0, 0)$, $(0, 0, 1)$, $(1, 1, 1)$ and $[0, 0, 1]$, $[1, 0, 0]$.

3. Prove 8.51 (Steiner's construction), using the projectively related lines $[\mu, 1, 0]$ and $[0, \mu, 1]$ through the fixed points $(0, 0, 1)$ and $(1, 0, 0)$.

4. Write down the envelope equations for

$$\text{(i) } x_2^2 + 2x_3x_1 = 0, \qquad \text{(ii) } x_2^2 = x_3x_1.$$

5. Use Exercise 7 of Section 12.2 for an algebraic solution to Exercise 1 of Section 8.5.

6. Given a conic $(xx) = 0$ and an exterior point (y) (see Section 8.1), describe, both algebraically and geometrically, the locus of a variable point (x) such that the line $(x)(y)$ is self-conjugate. Dualize this result.

7. If the four tangents through two exterior points are all distinct, their points of contact and the two exterior points all lie on a conic.

8. Given a conic $(xx) = 0$, let (z) be any exterior point. If (zz) does not happen to be a square, divide all the coefficients c_{ij} in (xx) by (zz). We now have the same conic expressed in such a way that (zz) *is* a square (namely, 1). When the equation $(xx) = 0$ has been thus normalized, any point (y) not on the conic is *exterior or interior according as* (yy) *is or is not a square*. (In real geometry this means, according as (yy) is positive or negative. In complex geometry we know already that there are no interior points because there are no elliptic involutions.) Illustrate this criterion by applying it to the conic (12.61).

9. In real geometry all interior points of a conic are mutually accessible, but in *rational* geometry the number of classes of mutually accessible points is infinite.

12.7 The Analytic Geometry of *PG*(2, 5)

It is proved in books on algebra that, if q is any power of a prime (such as 2, 3, 4, 5, 7, 8, 9, or 11), there is a field having just q elements. A famous theorem (Reference **15**, p. 94) tells us that a finite field must necessarily be commutative. We remarked, in Section 12.1, that our coordinates can belong to any such field, provided q is odd (that is, *not* a power of 2). This proviso comes from (12.32), where the determinant of the coordinates of A, B, C, being twice the determinant of the coordinates of P, Q, R, must be zero if $2 = 0$ (which is what happens in a field having 2^k elements).

We saw, in Section 12.2, that all the points on the general line $(y)(z)$ can be expressed in the form

$$(\mu y + z),$$

where μ runs over all the elements of the field and the extra element ∞, which yields (y). Hence the field with q elements (where q is any power of an odd prime) yields the finite projective plane with $q + 1$ points on each line, that is, in the notation of Section 10.1, $PG(2, q)$. (In an n-dimensional geometry such as $PG(n, q)$, a point has $n + 1$ coordinates.)

In $PG(2, q)$, the number of coordinate symbols (x_1, x_2, x_3), with q possible values for each x_i, is q^3. From this number we subtract 1, since the symbol $(0, 0, 0)$ has no geometric meaning. Moreover, each point (x) is the same as (λx) for $q - 1$ values of λ, namely all the nonzero elements of the field. In this way we see again that the number of points in the plane is

$$\frac{q^3 - 1}{q - 1} = q^2 + q + 1.$$

For each point (x) there is, of course, a corresponding line $[x]$; so the number of lines is the same.

When q is an odd prime (and not the square or higher power of such a prime), the elements of the field are simply the residue-classes modulo q. For instance, when $q = 5$, they may be denoted by the familiar symbols

$$0, 1, 2, 3, 4,$$

with the usual rules for addition and multiplication except that

$$\begin{aligned} 1 + 4 = 2 + 3 &= 0, & 2 + 4 = 3 + 3 &= 1, \\ 3 + 4 &= 2, & 4 + 4 &= 3, \\ 2 \cdot 3 = 4 \cdot 4 &= 1, & 3 \cdot 4 &= 2, \\ 2 \cdot 4 &= 3, & 3 \cdot 3 &= 4. \end{aligned}$$

These rules are quite natural, provided we interpret the symbols as follows: 0 means the set of all multiples of 5, in other words, all integers whose final digit is 0 or 5; 1 means the residue-class that includes all the positive integers whose final digit is 1 or 6, and so on. In a more familiar notation, such equations as

$$3 + 4 = 3 \cdot 4 = 2$$

are "congruences"

$$3 + 4 \equiv 3 \cdot 4 \equiv 2 \pmod{5}.$$

To assign coordinates to the 31 points P_r and 31 lines l_s of Section 10.2, we choose $P_1P_2P_{30}$ as triangle of reference, P_5 as unit point, and derive other points and lines in the manner of Figure 12.2A. The results are as follows:

$$
\begin{array}{ll}
P_0 = (0, 1, 2), & l_0 = [0, 1, 2], \\
P_1 = (1, 0, 0), & l_1 = [1, 0, 0], \\
P_2 = (0, 1, 0), & l_2 = [0, 1, 0], \\
P_3 = (1, 3, 1), & l_3 = [1, 3, 1], \\
P_4 = (1, 2, 0), & l_4 = [1, 2, 0], \\
P_5 = (1, 1, 1), & l_5 = [1, 1, 1], \\
P_6 = (2, 0, 1), & l_6 = [2, 0, 1], \\
P_7 = (0, 1, 4), & l_7 = [0, 1, 4], \\
P_8 = (3, 1, 2), & l_8 = [3, 1, 2], \\
P_9 = (2, 1, 0), & l_9 = [2, 1, 0], \\
P_{10} = (1, 0, 1), & l_{10} = [1, 0, 1], \\
P_{11} = (0, 1, 1), & l_{11} = [0, 1, 1], \\
P_{12} = (2, 3, 1), & l_{12} = [2, 3, 1], \\
P_{13} = (1, 4, 0), & l_{13} = [1, 4, 0], \\
P_{14} = (1, 2, 1), & l_{14} = [1, 2, 1], \\
P_{15} = (3, 2, 1), & l_{15} = [3, 2, 1], \\
P_{16} = (1, 0, 2), & l_{16} = [1, 0, 2], \\
P_{17} = (0, 2, 1), & l_{17} = [0, 2, 1], \\
P_{18} = (1, 1, 2), & l_{18} = [1, 1, 2], \\
P_{19} = (1, 1, 0), & l_{19} = [1, 1, 0], \\
P_{20} = (1, 4, 1), & l_{20} = [1, 4, 1], \\
P_{21} = (1, 1, 4), & l_{21} = [1, 1, 4], \\
P_{22} = (2, 1, 3), & l_{22} = [2, 1, 3], \\
P_{23} = (1, 3, 2), & l_{23} = [1, 3, 2], \\
P_{24} = (4, 1, 1), & l_{24} = [4, 1, 1], \\
P_{25} = (2, 1, 1), & l_{25} = [2, 1, 1], \\
P_{26} = (1, 1, 3), & l_{26} = [1, 1, 3], \\
P_{27} = (3, 1, 1), & l_{27} = [3, 1, 1], \\
P_{28} = (1, 2, 3), & l_{28} = [1, 2, 3], \\
P_{29} = (4, 0, 1), & l_{29} = [4, 0, 1], \\
P_{30} = (0, 0, 1), & l_{30} = [0, 0, 1].
\end{array}
$$

EXERCISES

1. Which points of $PG(2, 5)$ lie on the line [1, 0, 0]?
2. Which lines pass through the point (1, 1, 1)?

3. Find equations for the collineations

(i) $P_r \to P_{r+1}, \quad l_s \to l_{s-1}$;
(ii) $P_r \to P_{5r}, \quad l_s \to l_{5s+3}$.

4. Find equations for the polarity $P_r \leftrightarrow l_r$.

5. Which points lie on the conics
(i) $x_1^2 + x_2^2 + x_3^2 = 0$, (ii) $x_1^2 + x_2^2 - x_3^2 = 0$,
(iii) $x_2^2 = x_3x_1$, (iv) $x_2x_3 + x_3x_1 + x_1x_2 = 0$,
(v) $x_1^{\frac{1}{2}} \pm x_2^{\frac{1}{2}} \pm x_3^{\frac{1}{2}} = 0$?

[*Hint:* $2^{-1} = 3$, $3^{-1} = 2$, $4^{-1} = 4$.]

6. What condition must the coordinate field satisfy if there exists a configuration 8_3 (in the notation of Section 3.2) consisting of points A, B, C, D, E, F, G, H with the following triads collinear:

$$ABD,\ BCE,\ CDF,\ DEG,\ EFH,\ FGA,\ GHB,\ HAC?$$

When this condition is satisfied, the 8 points occur in 4 pairs of "opposites" whose joins AE, BF, CG, DH are concurrent, so that the complete figure is a $(9_4, 12_3)$. In the special case of the plane $PG(2, 3)$, this $(9_4, 12_3)$ is derived from the whole plane (which is a 13_4) by omitting one line and the four points that lie on it. [*Hint:* Take ABC as triangle of reference and

$$D = (1, 1, 0), \qquad E = (0, 1, 1), \qquad H = (1, 0, \omega).$$

Deduce $F = (1, 1, \omega + 1)$, $G = (1, \omega + 1, \omega)$. Obtain an equation for ω from the collinearity of FGA.]

7. In $PG(2, q)$, as in real geometry (Exercise 9 of Section 12.6), all interior points of a conic are mutually accessible. (We are still assuming q to be odd.)

12.8 Cartesian Coordinates

In Section 11.2 we saw how to derive affine space from projective space by specializing one plane, calling it "the plane at infinity," and then omitting it. The synthetic treatment required three dimensions, but the analytic treatment can be carried out just as easily in two dimensions. Accordingly, let us consider the *affine plane*, that is, the ordinary plane of elementary geometry, in which two lines are said to be *parallel* if they do not meet. We regard this as *part of* the projective plane, namely, the projective plane minus one line: "the line at infinity." Two lines are said to be parallel if they meet on this special line. We soon see that parallelism, so defined, has all its familiar properties. The apparent inconsistency, of saying that parallel lines meet and yet do not meet, is resolved by regarding the affine plane as being derived from the projective plane by *omitting* the special line (and all its points) while retaining the consequent concept of parallelism. This modification of projective geometry is called *affine geometry*.

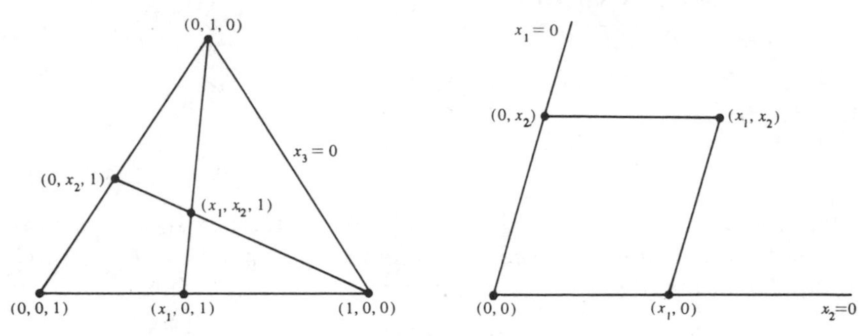

FIGURE 12.8A

When coordinates are used, it is convenient to take the line at infinity to be [0, 0, 1] or $x_3 = 0$, so that the points at infinity are just all the points (x) for which the third coordinate is zero. The points of the affine plane are thus all the points (x) for which the third coordinate is not zero. By a suitable multiplication (if necessary), any such point can be expressed in the form

$$(x_1, x_2, 1),$$

which can be shortened to (x_1, x_2). The two numbers x_1 and x_2 are called the *affine coordinates* of the point. In other words, if $x_3 \neq 0$, the point (x_1, x_2, x_3) of the projective plane can be regarded as the point $(x_1/x_3, x_2/x_3)$ of the affine plane; and the equation of any locus can be made into the corresponding equation in affine coordinates by setting $x_3 = 1$. In particular, the line $[X]$ has the equation

$$X_1x_1 + X_2x_2 + X_3 = 0,$$

and a pencil of parallel lines is obtained by fixing X_1 and X_2 (or, more precisely, the ratio $X_1 : X_2$) while allowing X_3 to take various values.

Let us see what has happened to the triangle of reference. Its first two sides have become the coordinate *axes*, namely the x_2-axis $x_1 = 0$ (along which x_2 varies) and the x_1-axis $x_2 = 0$ (along which x_1 varies). The third side, $x_3 = 0$, is the line at infinity. The first two vertices are the points at infinity on the axes: (1, 0, 0) on the x_1-axis and (0, 1, 0) on the x_2-axis. The third vertex is the *origin* (0, 0), where the two axes meet. (See Figure 12.8A.)

The homology (12.45) becomes a transformation of affine coordinates

(12.81) $$x'_1 = \mu x_1, \qquad x'_2 = \mu x_2$$

when we set $x_3 = x'_3 = 1$, which requires $\rho = \mu^{-1}$. Similarly, the elation (12.46), with $\rho = 1$, becomes

(12.82) $$x'_1 = x_1 + a_1, \qquad x'_2 = x_2 + a_2.$$

In either case [since both (12.47) and (12.48) involve $X'_1 = X_1$ and $X'_2 = X_2$], every line is transformed into a parallel line; in other words, directions are preserved. The homology leaves the origin invariant and multiplies the coordinates of every point by μ; we call this a *central dilatation*. The elation, leaving no (proper) point invariant, is a *translation* (or "parallel displacement"). These two affine transformations enable us to define relative distances along one line, or along parallel lines (Reference **7**, pp. 117–125); but affine geometry provides no comparison for distances in different directions.

A conic is called a *hyperbola*, a *parabola*, or an *ellipse*, according as the line at infinity is a secant, a tangent, or a nonsecant. (This agrees with the classical definitions, since a hyperbola "goes off to infinity" in two directions, a parabola in one direction, and an ellipse not at all.) The pole of the line at infinity is called the *center* of the conic. In the case of a hyperbola, this is an exterior point (see Section 8.1), and the two tangents that can be drawn from it are the *asymptotes*, whose points of contact are at infinity. Thus

$$x_1x_2 = 1 \quad \text{and} \quad x_1^2 - x_2^2 = 1$$

are hyperbolas, $x_2^2 = x_1$ is a parabola, and in *real* geometry,

(12.83) $$x_1^2 + x_2^2 = 1$$

is an ellipse.

To pass from affine geometry to *Euclidean geometry* we select, among all the ellipses centered at the origin, a particular one, and call it the *unit circle*. This provides units of measurement in all directions. To pass from affine coordinates to *Cartesian coordinates* we choose, as unit circle, the ellipse (12.83). The dilatation (12.81) transforms this into a circle of radius μ:

(12.84) $$x_1^2 + x_2^2 = \mu^2.$$

The translation (12.82) then yields the general circle

$$(x_1 - a_1)^2 + (x_2 - a_2)^2 = \mu^2.$$

The rest of the story is told in every textbook on analytic geometry.

EXERCISES

1. Give equations (in affine or Cartesian coordinates) for
   (i) the half-turn about the origin;
   (ii) the half-turn about the point (a_1, a_2). (See Exercise 4 of Section 6.3.)

2. Describe the collineation (in Cartesian coordinates)

$$x'_1 = x_1 \cos \alpha - x_2 \sin \alpha, \; x'_2 = x_1 \sin \alpha + x_2 \cos \alpha.$$

[*Hint:* Consider its effect on the circle (12.84).]

3. Give equations (in Cartesian coordinates) for
   (i) the reflection in the x_1-axis;
   (ii) the reflection in the line $x_1 = x_2$.

12.9 Planes of characteristic two

We noticed, in Section 12.7, that Axiom 2.17 excludes coordinate fields in which $1 + 1 = 0$. From the algebraic standpoint this exclusion seems artificial, so it is interesting to see what happens if we replace this axiom by its opposite:

12.91 *The three diagonal points of a quadrangle are always collinear.*

The principle of duality still holds, also the fundamental theorem and the basic properties of quadrangular sets. But there are no harmonic sets: in Figure 2.5A, the points C and F coincide! A projectivity relating ranges on two distinct lines still has an axis, but in the case of a perspectivity the axis passes through the center. (In Figure 4.3A on page 37, EN coincides with EO.) In contrast with 5.41, *a one-dimensional projectivity is parabolic if and only if it is an involution.* But elliptic and hyperbolic projectivities of odd period are still possible. In contrast with 6.32, *every projective collineation of period* 2 *is an elation.* As a conic is no longer a self-dual concept, von Staudt's definition (page 72) has to be replaced by Steiner's (page 80). Related pencils yield a conic locus, while related ranges yield a conic envelope, which is quite different. A conic locus has a *nucleus:* a point that lies on every polar and, in particular, on every tangent! For "pseudo-polarity," see Reference **15**, pp. 242–245.

Answers To Exercises

Section **1.2**

1. (iv), (vi), (vii), (viii).

2. So that just one "ray" (from the lamp to some point on the rim) is parallel to the wall.

3. (i) An ordinary point and a point at infinity are joined by just one line.
   (ii) Each ordinary line contains just one point at infinity.

Section **1.3**

2. Yes. Such apparently paradoxical behavior is characteristic of infinite sets.

Section **1.4**

(i) The configuration formed by n points (no 3 collinear) and the

$$\binom{n}{2} = \tfrac{1}{2}n(n-1)$$

lines that join them in pairs.

(ii) The configuration formed by n lines (no 3 concurrent) and their $\binom{n}{2}$ points of intersection.

Section **1.5**

1. So that distinct points X will yield distinct lines x.

3. In both these special cases, X'' coincides with X.

Section **1.6**

1. In the notation of Figure 1.6E, $ABC \overset{P}{\barwedge} SRC \overset{Q}{\barwedge} BAC$.
2. In the notation of Figure 3.3A, $abc \overset{p}{\barwedge} src \overset{q}{\barwedge} bac$.
3. Use the method of Figure 1.6B, taking A'', B'', C'' to be three collinear points on a, b, c, respectively.
4. Proceed as in Figure 1.6D with the names of the points B and D interchanged.

Section **2.1**

1. (i) Axioms 2.11 and 2.12 yield one point and three others.
   (ii) If this statement were not true, all the points that exist would lie on a. By Axiom 2.12, every line would contain two (in fact, three) of these points. By Axiom 2.13, every line would coincide with a, that is, a would be the only line, contradicting Axiom 2.11.
   (iii) Let P and l be the point and line described in Axiom 2.11. If A is not on l, l fulfills our requirement. If A is on l, Axiom 2.12 yields another point B on l. By Axiom 2.13, a line is equally well determined by any distinct two of its points. The line $l = AB$, not containing P, is distinct from BP. Therefore BP is a line not passing through A.
   (iv) Consider the point A. By Axiom 2.12, the line whose existence is proved in (iii) contains three distinct points, each of which (by Axiom 2.13) can be joined to A by a line. The three lines so obtained are distinct; for, if two of them, AB and AC, were coincident, the same line could be called BC, contradicting the fact that BC does not pass through A.
2. In the notation of Figure 5.1A,
$$AECF \overset{P}{\barwedge} SRCU \overset{Q}{\barwedge} BDCF.$$
   This projectivity $AECF \barwedge BDCF$ has C and F as invariant points. By Axiom 2.18, it has no other invariant point.
3. 2.17.

Section **2.2**

1. (i) By 2.24, there exists a quadrangle $PQRS$ having 6 distinct sides
$$p = PQ, \quad q = PS, \quad r = RS, \quad s = QR, \quad u = QS, \quad w = PR.$$
   The first 4 of these 6 lines form a quadrilateral $pqrs$ having 6 distinct vertices
$$P = p \cdot q, \quad Q = p \cdot s, \quad R = r \cdot s, \quad S = q \cdot r, \quad U = q \cdot s, \quad W = p \cdot r$$
   (Reference **8**, pp. 231–232).

(ii) If there exists a quadrilateral $pqrs$ whose diagonal lines

$$u = QS, \quad v = UW, \quad w = PR$$

are concurrent, it follows that the point $u \cdot w$ lies on v, contradicting Axiom 2.17 (which tells that the diagonal points

$$U = q \cdot s, \quad V = u \cdot w, \quad W = p \cdot r$$

of the quadrangle $PQRS$ are not collinear).

Section **2.3**

1. We have $O = PP' \cdot QQ'$; R' may be anywhere on the line OR.

3. If the triangles are congruent by translation, they are perspective from a point at infinity. If they are merely homothetic (differing in size, but still similar and similarly situated) they are perspective from an ordinary point.

Section **2.4**

1. If AP, BQ, CR are concurrent, as in Figure 2.4A, we have immediately $(AD)(BE)(CF)$. Conversely, if $(AD)(BE)(CF)$ while the lines AP, BQ, CR are not known to be concurrent, define $S = AP \cdot BQ$, $C' = AB \cdot RS$. The quadrangle $PQRS$ yields $(AD)(BE)(C'F)$. By Theorem 2.41, $C' = C$. Therefore CR passes through S.

2. Let the diagonal triangles be LMN and $L'M'N'$, so that $L = PS \cdot QR, \ldots,$ $L' = P'S' \cdot Q'R', \ldots.$ The triangles PQL and $P'Q'L'$ are perspective from g, since pairs of corresponding sides meet in D, A, F. Therefore they are perspective from $O = PP' \cdot QQ'$. Thus O lies on LL', and similarly on MM', NN'.

Section **2.5**

1. $\mathrm{H}(PQ, FG)$, $\mathrm{H}(RS, CG)$.

2. As in Figure 1.6E.

3. Of course, the concept of a point "inside" a triangle does not belong to projective geometry. But what happens in the Euclidean plane is that the line AB separates R from the other three vertices of the quadrangle.

4. Since F coincides with C, the four points reduce to three. Each of the three is its own harmonic conjugate with respect to the other two!

5. C is the midpoint of AB.

6. Measuring from O, we have the distances 10, 12, 15 in "harmonic progression." (Their reciprocals are in "arithmetic progression.") Measuring from C, we have another harmonic progression 3, 5, 15.

Section **3.1**

2. Certain pairs of triangles are perspective from points P, Q, R, S, and consequently also from lines p, q, r, s. For instance, ABC and SRQ are perspective from P.

Section **3.2**

2. See the frontispiece.
4. No, not in the projective geometry that we have defined. A 7_3 would consist of a quadrangle with collinear diagonal points, as in Exercise 3 of Section 2.1.

Section **3.3**

1. We construct, in turn, $A_2 = BC \cdot AS$, $B_2 = CA \cdot BS$, $C_1 = AB \cdot A_2B_2$, $P = AS \cdot CC_1$, $Q = BS \cdot CC_1$, $R = CS \cdot AQ$. (Compare Exercise 2 of Section 3.1.)
2. The quadrangle BCB_1C_1 tells us that CC_1 meets AA_1 in the harmonic conjugate of R (with respect to A and A_1), which is Q. Similarly, the quadrangle CAC_1A_1 tells us that CC_1 meets BB_1 in the harmonic conjugate of R, which is P. Finally, the quadrangle ABA_1B_1 tells us that Q (on AA_1) and P (on BB_1) are harmonic conjugates with respect to C and C_1.

Section **3.4**

1. The trilinear polar of a vertex is indeterminate: any line through this vertex will serve. The trilinear polar of any other point on a side is that side itself. The trilinear pole of a side is indeterminate: any point on this side will serve. The trilinear pole of any other line through a vertex is that vertex itself.
2. S.
3. $a = QR$, $b = RP$, $c = PQ$, $d = AP$, $e = BQ$, $f = CR$, $p = BC$, $q = CA$, $r = AB$, $s = DE$.
4. The three medians are concurrent.

Section **3.5**

1. Yes; M is the harmonic conjugate of B with respect to A and C.
2. Yes, since $\mathrm{H}(B'M, A'C')$, $\mathrm{H}(C'M, B'D')$, ...
3. The points of the sequence are evenly spaced along the line (so that the segments AB, BC, ..., are all congruent).

Section **4.1**

1. Choose the arbitrary points, E, F, and construct $O_1 = AE \cdot BF$, $O_2 = BE \cdot CF$, $O_3 = CE \cdot AF$, $D = CE \cdot BF$. Then

$$ABC \overset{O_1}{\bar{\wedge}} EDC \overset{O_2}{\bar{\wedge}} BDF \overset{O_3}{\bar{\wedge}} BCA.$$

2. Suppose F is transformed into G. Applying the same projectivity three times, we have

$$ABCD \;\bar{\wedge}\; BCAE \;\bar{\wedge}\; CABF \;\bar{\wedge}\; ABCG.$$

By Axiom 2.18, $G = D$. In other words, this is a projectivity *of period three.*

3. Theorem 1.63 tells us that any "double interchange," $(AB)(CD)$ or $(AC)(BD)$ or $(AD)(BC)$, can be effected by a projectivity.

Section **4.2**

1. Any point K of the harmonic net $\mathsf{R}(ABC)$ (or of the harmonic sequence $ABC\ldots$) is derived from ABC by a finite sequence of harmonic constructions, each of which is carried over by the projectivity $ABC \;\bar{\wedge}\; A'B'C'$. Hence K is transformed into a corresponding point K' of the harmonic net $\mathsf{R}(A'B'C')$ (or of the harmonic sequence $A'B'C'\ldots$).

2. These three points

$$Q = PC \cdot A_1A, \quad C_2 = CS \cdot AB, \quad D = SA_1 \cdot BP$$

are *collinear*, since $\mathsf{H}(RS, CC_2)$, $\mathsf{H}(RB, PD)$, and $RSC \overset{Q}{\bar{\wedge}} RBP$.

3. A projectivity relating pencils through two distinct points is a perspectivity if and only if it has an invariant line.

4. Use 4.21.

Section **4.3**

Every projectivity relating pencils through two distinct points determines another special point, the "center," which lies on the join of the cross-intersections of any two pairs of corresponding lines.

Section **4.4**

1.

a_1	b_1	c_1	a_2	b_2	c_2	a_3	b_3	c_3
A_3	A_1	A_2	A_2	A_3	A_1	A_1	A_2	A_3
B_1	B_3	B_2	B_3	B_2	B_1	B_2	B_1	B_3
C_2	C_2	C_2	C_1	C_1	C_1	C_3	C_3	C_3

2. Construct the points $C_1 = A_1B_1 \cdot A_3B_2$, $C_2 = A_2B_2 \cdot A_3B_1$ (Figure 4.4B) and then $B_3 = A_2C_1 \cdot A_1C_2$. The concurrence of A_1B_2, A_2B_1, A_3B_3 follows from 4.41, applied to the hexagon $A_1B_2C_1A_2B_1C_2$.

3. The diagonals A_1B_2, B_3A_3, A_2B_1 all pass through C_3.

4. If the six sides of a plane hexagon pass alternately through two points, the three diagonals are concurrent. One such hexagon is $a_1b_2c_1a_2b_1c_2$.

5. No. The hexagon $ABPCSA_1$ is very special, as $APS \barwedge A_1BC$ and $APS \barwedge A_1CB$.

6. Suppose $A_1B_1C_2$ is inscribed in $A_2B_2C_3$, as in Figure 4.4B. Any point C_1 on A_1B_1 determines two further points

$$A_3 = B_1C_2 \cdot B_2C_1, \qquad B_3 = C_1A_2 \cdot C_2A_1,$$

which, by Pappus's theorem, are collinear with C_3.

7. $8 = A$, $0 = B$, $1 = C$, $2 = A'$, $3 = B'$, $4 = C'$, $5 = L$, $6 = M$, $7 = N$. This notation has the advantage that the incidences are maintained when we replace each number x by either $2x$ or $x + 3$ and reduce modulo 9. Applying the former transformation to the given triads of collinear points, we obtain the whole set of nine triads:

801, 234, 567,
702, 468, 135,
504, 837, 261.

Applying it to the given cycle of Graves triangles, we obtain the whole set of six cycles:

012, 345, 678,
024, 681, 357,
048, 372, 615,
087, 654, 321,
075, 318, 642,
051, 627, 384.

Section 5.1

1. $ADBF \overset{Q}{\barwedge} AVSP \overset{R}{\barwedge} ADCE.$

3. The line joining the centers of the two perspectivities would inevitably yield an invariant point (such as F in Figure 5.1A, or C in Figure 5.1B).

4. Our axioms (not being categorical) provide no answer to this question. In *real* geometry, the projectivity $ABC \barwedge BCA$ (Exercise 2 of Section 4.1) provides an instance of an elliptic projectivity. But in *complex* geometry every projectivity is either hyperbolic or parabolic (because every quadratic equation can be solved).

Section **5.2**

The parabolic projectivity becomes a *translation*, shifting every point along the line through the same distance AA'. The point at infinity, C, is evidently the only invariant point.

Section **5.3**

1. This is merely 5.34 in a more symmetrical notation. The remarks about Desargues at the beginning of Section 5.3 suggest an analogous theorem concerning six *numbers* a, b, c, d, e, f (which we may think of as distances of the points from an arbitrary "origin"):

   If $a + b = d + e$ and $b + c = e + f$, then $c + d = f + a$.

2. The product leaves both M and N invariant, and transforms A into A'.

3. Watch what happens to A, A', B.

4. Let A be any noninvariant point. The given projectivity, not being an involution, can be expressed as

   $$AA'A'' \barwedge A'A''A''',$$

   where the three points on either side are all distinct. (Conceivably A''' coincides with A, and then the second involution, like the first, is expressed in terms of one of its invariant points.)

5. The symbols $(AB')(BA)$ and $(AB)(BA')$ are meaningless. However, the projectivity $ABC \barwedge ABC'$ is the product of $(AB)(CC)$ and $(AB)(CC')$.

Section **5.4**

1. The given statement says that C and D are the invariant points of an involution that interchanges A and B.

2. If $\mathsf{H}(AB, MN)$, the fundamental theorem serves to identify the hyperbolic projectivity $AMN \barwedge BMN$ with the involution $(MM)(NN)$.

3. The involution $(MM)(NN)$ has the effect $AB'MN \barwedge BA'MN$, which shows that MN is a pair of the involution $(AA')(BB')$.

4. $A'E'C'F' \barwedge AECF \barwedge BDCF \barwedge B'D'C'F'$.

5. Since the involution $ABCD \barwedge BAED$ has CE for one of its pairs, its invariant points are D and D'. Therefore $\mathsf{H}(AB, DD')$.

6. The hint shows that the projectivity $X \barwedge X'$ has *only one* invariant point, namely B.

7. By 4.21, there is a projectivity $BCAD \barwedge ACBE$. Since A and B are interchanged, 5.31 shows that this projectivity is an involution. Since C is one invariant point, 5.41 shows that F is the other. Since DE is a pair of this involution, $\mathrm{H}(DE, CF)$. By 4.21 again, there is a projectivity $ABCF \barwedge DEFC$. By 5.31 again, this is an involution. By 5.33, it follows that $(AD)(BE)(CF)$.

Section **6.1**

(i) This is a collineation of period 2, transforming the side PQ into itself according to the hyperbolic involution $(PQ)(UU)$, where $U = PQ \cdot RS$. On the opposite side RS, every point is invariant.

(ii) This is a collineation of period 3, transforming the trilinear polar of S into itself according to the projectivity $DEF \barwedge EFD$.

(iii) This is a collineation of period 4, leaving invariant the point $PR \cdot QS$ and the line $(QR \cdot PS)(PQ \cdot RS)$.

Section **6.2**

1. The center is $PP' \cdot QQ'$ and the axis joins D to $PQ \cdot P'Q'$.

2. Hyperbolic or parabolic, according as the collineation is a homology or an elation.

3. Since $ABOO \barwedge A'B'OO$, we have $(AB')(A'B)(OO)$.

4. Clearly, each point on the axis o is invariant. If any other point A were invariant too, the first elation would take A to some different point A', and the second would take A' back to A: the two elations would be $[o; A \to A']$ and $[o; A' \to A]$, whose product is the identity. Apart from this trivial case, the product can only be an *elation* (not a homology). In fact,

$$[o; A \to B][o; B \to C] = [o; A \to C].$$

Alternatively, the dual result can be proved as follows: If two elations have the same center O, they induce parabolic projectivities on any given line through O. By 5.21, their product, inducing another parabolic projectivity on this line, must itself be an elation (if it is not merely the identity).

5. Let A and B be the centers of two such elations, α and β, which transform P into P^α and P^β. If A and B are distinct, and P is any point not on the axis AB, the point $AP^\beta \cdot BP^\alpha$ is both $P^{\alpha\beta}$ and $P^{\beta\alpha}$; therefore $\alpha\beta = \beta\alpha$. If A and B coincide, let γ be an elation having a different center (but the same axis), so that α commutes with both γ and $\beta\gamma$. Then $\alpha\beta\gamma = \beta\gamma\alpha = \beta\alpha\gamma$; therefore $\alpha\beta = \beta\alpha$.

6. A perspective collineation; namely, an elation or a homology according as the join of the points and the intersection of the lines are or are not incident.

7. If four lines do not form a quadrilateral, they can be described as three lines through a point O and a fourth line o. These are, of course, invariant for a perspective collineation with center O and axis o.

8. A *dilatation* (Reference **8**, p. 68); more specifically, a *translation* or a *central dilatation* (like a photographic enlargement) according as the center is a point at infinity or an ordinary point, that is, according as the perspective collineation is an elation or a homology. An elation is called a *shear* if its axis is an ordinary line while its center is the point at infinity on this line.

Section **6.3**

1. An elation. In fact, since the product is evidently a perspective collineation having the same axis, we merely have to eliminate the possibility of a homology. For this purpose we apply, to the line joining the two centers, Exercise 6 of Section 5.4, which tells us that the product of two hyperbolic involutions with a common invariant point is a *parabolic* projectivity.
2. In the notation of Figure 6.2A, the elation $[o; P \to P']$ may be expressed as the product of harmonic homologies with centers P and O_1, O_1 being the harmonic conjugate of O with respect to P and P'. Knowing, from Exercise 1, that this product is some kind of elation, we merely have to observe that the first homology leaves P invariant and the second takes P to P'.
3. The identity. In fact, choosing four points P, Q, R, S, on the first two sides of the given triangle ABC, in such a way that $\mathrm{H}(BC, PQ)$ and $\mathrm{H}(CA, RS)$, we find $$PQRS \to PQSR \to QPSR \to PQRS.$$
4. Let O_1, O_2, o_1, o_2 be the centers and axes of two harmonic homologies whose product is a homology with center O and axis o. By Exercise 1, $o_1 \neq o_2$ (for, if o_1 and o_2 were the same, the product would be an elation, not a homology). Dually, $O_1 \neq O_2$. Any point P on o, being invariant for the product, is interchanged with the same point P' by both the harmonic homologies. Hence O_1 and O_2 are collinear with P and P'. Since P is arbitrary on o, it follows that $o = O_1O_2$. Dually, $O = o_1 \cdot o_2$. Both the harmonic homologies induce on o an involution $P \barwedge P'$ whose invariant points are O_1 and $o \cdot o_1$, also O_2 and $o \cdot o_2$. But $O_1 \neq O_2$. Therefore $O_1 = o \cdot o_2$ and $O_2 = o \cdot o_1$. Finally, by applying Exercise 3 to the triangle O_1O_2O, we see that the product of the harmonic homologies with centers O_1, O_2 and axes o_1, o_2 is the harmonic homology with center O and axis o.

5. A *half-turn* (or rotation through 180°, or reflection in a point). Similarly, the reflection in a line is a harmonic homology having the line as axis while its center is at infinity in the perpendicular direction.

Section **6.4**

Each diagonal point of the quadrangle is determined by two opposite sides, and these are transformed into two opposite vertices of the quadrilateral.

Section **7.1**

1. A self-conjugate line contains its pole. By 7.11, it cannot contain any other self-conjugate point.
2. $Y = x \cdot b$.
4. $r \cdot BC$ and $s \cdot BC$.
5. No; a trilinear polarity is not a correlation, as it transforms collinear points into non-concurrent lines. Moreover, it is not a one-to-one correspondence, as it transforms any point on a side of the triangle into that side itself.

Section **7.2**

1. Such a correlation induces on g the involution $DEF \barwedge ABC$. Thus we can apply 7.21 to the triangle SAD (or SBE, or SCF).
2. Since P and p are self-conjugate, 7.11 shows that Q may be any other point on p (but not on a side of the triangle ABC). For instance, we may take

$$Q = p \cdot (a \cdot AP)(b \cdot BP), \qquad q = P[A(a \cdot p) \cdot B(b \cdot p)].$$

Section **7.3**

1. The pole of the line $(p \cdot QR)(q \cdot RP)$ is the point $P(q \cdot r) \cdot Q(r \cdot p)$.
2. The polarity is $(ABC)(Pp)$, where $C = a \cdot b$.

Section **7.4**

1. $(a \cdot p)[A(p \cdot x) \cdot P(a \cdot x)] \cdot (b \cdot p)[B(p \cdot x) \cdot P(b \cdot x)]$.
2. The pole of such a line WX is

$$B[AP \cdot (a \cdot PW)(p \cdot b)] \cdot C[AP \cdot (a \cdot PX)(p \cdot c)].$$

Section **7.5**

1. Each vertex is the pole of the opposite side.
2. Use Exercise 1 of Section 7.4, with b, AP, BP, a replaced by q, r, s, t.

Section **7.6**

Apply the dual of Hesse's theorem to the quadrangle $RSUV$, in which $RS = p$ is conjugate to UV (through P) and $RV = t$ is conjugate to SU (through T).

Section **7.7**

1. Yes. In the notation of Section 6.1, Exercise (iii), the projective collineation $PQRS \to QRSP$ interchanges $PQ \cdot RS$ and $QR \cdot SP$.

2. The projective collineation $ABCP \to ABCP'$ (where neither P nor P' is on a side of the triangle ABC) is the product of the two polarities

$$(ABC)(Pp) \quad \text{and} \quad (ABC)(P'p),$$

where p is any line not through a vertex.

Section **8.1**

1. Every point on any line is conjugate to the pole of the line.

2. The polar of $p \cdot q$ is PQ.

3. Yes. Any two tangents meet in a point that is exterior.

4. Yes. If an interior point exists, its polar exists and is neither a tangent nor a secant.

5. PQR and pqr are polar triangles.

6. Consider the vertices on the side r of the circumscribed triangle pqr. Apply 8.13 to the conjugate lines $(p \cdot q)\ R$ and $(p \cdot q)(r \cdot PQ)$.

7. (i) an ellipse (being finite),
 (ii) a parabola (extending to infinity in one direction),
 (iii) a hyperbola (extending to infinity in two directions),
 (iv) the center,
 (v) the asymptotes,
 (vi) a diameter (bisecting a set of parallel chords),
 (vii) conjugate diameters.
 [For a complete account, see Reference **7**, pp. 129–159.]

8. The line CD, joining the midpoint of the chord PQ to the pole of the line PQ, is a diameter. Its two intersections with the parabola are the center (at infinity) and S.

Section **8.2**

1. By 8.21, QR meets SP in a point conjugate to B. But the only such point on QR is A.

2. Let AP meet the conic again in S. Then the diagonal triangle of the quadrangle $PQRS$ has the desired properties.

3. Let BQ and BR meet the conic again in S and P, respectively. By Exercise 1, A and B are two of the diagonal points of the quadrangle $PQRS$. The third diagonal point C lies on the side PQ, as in Figure 8.2A.

4. Let C and AB be the center and axis of such a homology. In the notation of Figure 8.2A, $\mathrm{H}(CC_1, PQ)$. Since every secant PQ through C meets AB in the harmonic conjugate of C, the homology interchanges P and Q.

5. The inscribed quadrangle $PQRS$ and the circumscribed quadrilateral $pqrs$ have the *same* diagonal triangle.

6. In terms of $A' = PD \cdot BC$, the polarity $(ABC)(Dd)$ induces, on PD, the involution $(AA')(DP')$, which is $(PP)(SS)$. Therefore P and S are self-conjugate. By Exercise 3 (applied to PS instead of QR), Q and R also are self-conjugate.

Section **8.3**

1. If a triangle is circumscribed about a conic, any point conjugate to one vertex is joined to the other two vertices by conjugate lines. Let a variable tangent of a conic meet two fixed tangents in X and Y; then $X \barwedge Y$.

2. If we regard PT, PE, PA, PU as four positions of x, the corresponding positions of y are QT, QA, QB, QU.

Section **8.4**

2. Any line h through R determines C, C_1, c, C_2, and finally S, the harmonic conjugate of R with respect to C and C_2. Accordingly, we construct c by joining D to $P(q \cdot h) \cdot Q(p \cdot h)$. Then S is the point where h meets either of the lines
$$P(c \cdot QR), \qquad Q(c \cdot PR).$$

3. Dualizing Exercise 2, let p, q, r be the given tangents, P and Q the points of contact of p and q, and H any point on r. Let PQ meet $(p \cdot QH)(q \cdot PH)$ in C. Then another tangent s joins H to either of the points
$$p \cdot C(q \cdot r), \qquad q \cdot C(p \cdot r).$$

4. Let P, Q, R be three points on the first conic, and P', Q', R' (or p', q', r') the corresponding points (or tangents) of the second. Let D be the pole of PQ for the first conic, D' the pole of $P'Q'$ for the second (and d' the polar of $p' \cdot q'$). In view of Exercises 2 and 3, we merely have to relate the quadrangle $PQRD$ by a projective collineation (or correlation) to the quadrangle $P'Q'R'D'$ (or to the quadrilateral $p'q'r'd'$).

5. Retaining P, p, Q, q, but letting C vary on PQ, we observe that any such point has the same polar for all the conics, namely DC_1, where $D = p \cdot q$ and $\mathsf{H}(PQ, CC_1)$. Exercise 4 of Section 8.2 shows that each conic is transformed into itself by the harmonic homology with center C and axis c.

Section **8.5**

1. Let $s = DE$ be the variable line through O. After using the hint, we observe that, by 8.51 with $x = PA$ and $y = QB$, the locus of $S = x \cdot y$ is a conic through P, Q, R.

2. Defining $y = QX$, we have $x \barwedge X \barwedge y$. But PQ is an invariant line of the projectivity $x \barwedge y$. Hence $x \doublebarwedge y$, and the locus is a line.

3. The line joins the points of contact of the remaining tangents from P and Q.

4. By comparing ranges on RP' and $Q'R$, we see that, if $y = QL$,

$$x \barwedge M \overset{N}{\doublebarwedge} L \barwedge y.$$

Therefore the locus of $R' = x \cdot y$ is a conic.

Section **9.1**

1. AB, p, q are three tangents; P, Q are the points of contact of the last two. Each point Z on PQ yields another tangent XY.

2. If EA, AB, BC, CD, DE are the five tangents, the point of contact of the last lies on the line $B(AD \cdot CE)$.

3. A parabola (since $X_\infty Y_\infty$ is the line at infinity).

Section **9.2**

1. Dualizing Exercise 2 of Section 9.1, we see that the tangent at P to the conic $PQRST$ joins P to the point

$$RS \cdot (TP \cdot QR)(ST \cdot PQ).$$

2. An arbitrary line through P, say z, meets the conic again where it meets the line

$$S[(PQ \cdot s)(RS \cdot z) \cdot QR].$$

3. $\frac{1}{2}5! = 60$.

4. $KPQVON$ (or $KNOVQP$).

5. Consider the dissection of the two triangles RAP and RCQ (Figure 11.1A) into three quadrilaterals (each) by means of the perpendiculars x, y, z from M to RA, AP, PR; and x', y', z' from N to RC, CQ, QR. (The same symbols will serve for the segments and their lengths.) Comparison of

angles shows that the quadrilateral Axy is directly similar to $Cy'x'$, and Pyz to $Qz'y'$. Therefore

$$\frac{x}{y} = \frac{y'}{x'}, \qquad \frac{y}{z} = \frac{z'}{y'}, \qquad xx' = yy' = zz',$$

and

$$\frac{x}{z} = \frac{z'}{x'}.$$

Hence the quadrilateral Rxz is directly similar to $Rz'x'$, and the diagonal MR of the former is along the same line as the diagonal RN of the latter. (See also H. G. Forder, *Higher Course Geometry*, Cambridge University Press, 1949, p. 13.)

Section **9.3**

1. The desired point lies on the line

$$T[QR \cdot (PQ \cdot ST)(g \cdot RS)].$$

2. Of the conics through P, Q, R, S (Figure 9.3A), any one that touches g does so at an invariant point of Desargues's involution.

3. In this case, A is an invariant point of Desargues's involution. Yes, $TEAU \barwedge TABU \barwedge UBAT$.

4. Since t is a tangent, Desargues's involution on it is hyperbolic. Another conic arises from the second invariant point.

Section **9.4**

1. $\frac{1}{2}\binom{6}{3} = 10.$

2. This follows from 9.41 along with the latter half of 9.42.

3. Let PQR (Figure 9.4A) be the given triangle, and ST any secant of the former conic that is a tangent of the latter. Let the remaining tangents from S and T meet in U. By the dual of Exercise 2, U lies on the conic $PQRST$.

4. $PQRS$ and $PQRT$ are two quadrangles inscribed in the conic $PQRST$. By 8.21, their diagonal triangles, ABC and $A'B'C'$, are self-polar. By 9.42, the six vertices of these two triangles lie on a conic.

Section **9.5**

1. With respect to a degenerate conic consisting of two lines a and b, the polar of a general point P is the harmonic conjugate of $(a \cdot b)P$ with respect to a and b; the polar of a general point on a is a itself; the polar of $a \cdot b$ is

indeterminate; the pole of a general line is $a \cdot b$; the pole of a general line through $a \cdot b$ is any point on the harmonic conjugate line; and the pole of a is any point on a. The other kind of degenerate conic has the dual properties.

2. The other conic may be degenerate.

Section **10.1**

1. $q^2 + q + 1$; $(q^2 + 1)(q^2 + q + 1)$.
2. $q^3(q + 1)(q^2 + q + 1)/6$.
3. $n = q^2 + q + 1$, $d = q + 1$.

Section **10.2**

2.

r	12	11	10	9	8	7	6	5	4	3	2	1	0
s	1	2	3	4	5	6	7	8	9	10	11	12	0
	2	3	4	5	6	7	8	9	10	11	12	0	1
	4	5	6	7	8	9	10	11	12	0	1	2	3
	10	11	12	0	1	2	3	4	5	6	7	8	9

Section **10.3**

1.

r	6	5	4	3	2	1	0
	1	2	3	4	5	6	0
s	2	3	4	5	6	0	1
	4	5	6	0	1	2	3

The diagonal points of the quadrangle $P_2P_4P_5P_6$ are the three points on l_0.

Section **10.4**

1. 24; 9; 3. If A, B, C, D are the four points on a line, the three elliptic involutions are

$$(BC)(AD), \qquad (CA)(BD), \qquad (AB)(CD).$$

2. By considering the possible effect of a given involution on three distinct points A, B, C, we see that there are q involutions $(AA)(BX)$, where $X \neq A$, $q-1$ involutions $(AB)(CY)$, where $Y \neq A$ and $Y \neq B$, and $(q-1)^2$ involutions $(AY)(BZ)$, where $Z \neq A$ and $Z \neq Y$: altogether

$$q + (q - 1) + (q - 1)^2 = q^2$$

involutions. The number of hyperbolic involutions, such as $(AA)(BB)$, is obviously $\binom{q+1}{2}$; therefore the remaining $\binom{q}{2}$ are elliptic.

3. This follows from 2.51.

Section **10.5**

1. $l_s \to l_{5s+3}$. In fact, if $r + s \equiv 0, 1, 3, 8, 12$, or $18 \pmod{31}$,
$$5r + (5s + 3) = 5(r + s) + 3 \equiv 3, 8, 18, 12, 1 \text{ or } 0.$$

2. $13 \cdot 12 \cdot 9 \cdot 4 = 5616;\ 13 \cdot 9 = 117.$

3. $q + 1;\ q + 2.$

Section **10.6**

1. Since this collineation has period 3, Figure 7.7B must be modified so that $l''' = l$ and therefore $C = B''$, $C' = B$. Choosing $A = P_{10}$ and $l = l_2$, we obtain the polarities
$$(P_{10}P_2P_6)(P_{19}l_{13}), \qquad (P_{19}P_2P_{26})(P_{10}l_2).$$

2. Since 5 of the 6 points on a conic can be chosen in 6 ways, the number of conics is
$$\frac{31 \cdot 30 \cdot 25 \cdot 16 \cdot 6}{5!\,6} = 3100$$
(Reference 15, p. 267).

3. $\dfrac{16 \cdot 6}{3!} = 16;\ \dbinom{6}{3} = 20;\ \dfrac{3875 \cdot 16}{20} = 3100.$

4. 234. In $PG(2, 3)$, any 5 points include 3 that are collinear; thus 9.21 holds vacuously. In Section 9.5 we saw that any 5 points, no 4 collinear, determine a conic (possibly degenerate). This result remains significant in $PG(2, 3)$, but the conic is necessarily degenerate.

5. Yes. The polarity $P_r \leftrightarrow l_r$ determines the conic $P_0P_7P_8P_{11}$ which, regarded as a quadrangle, has the diagonal triangle $P_4P_{10}P_{12}$. The sides of this triangle (namely, l_4, l_{10}, l_{12}) are nonsecants: the points on them are just all the points in the plane *except* those that form the conic.

6. Let C be any point on the polar of P. Then CP is a pair of conjugate points and $\mathrm{H}(CP, PP)$; therefore P is accessible from itself. Since the relation $\mathrm{H}(CC_1, PQ)$ implies $\mathrm{H}(CC_1, QP)$, accessibility is symmetric. To prove that accessibility is transitive, let R be accessible from Q on one line and from P on another, so that $\mathrm{H}(AA_1, QR)$ and $\mathrm{H}(BB_1, RP)$, where AA_1 and BB_1 are suitable pairs of conjugate points. By Exercise 2 of Section 3.3, P and Q are harmonic conjugates with respect to the two points
$$C = AB_1 \cdot BA_1 \quad \text{and} \quad C_1 = AB \cdot A_1B_1$$
which, being opposite vertices of the quadrilateral $ABCA_1B_1C_1$, are conjugate (by 7.61: Hesse's theorem).

7. Let Q be accessible from P; that is, let $\mathrm{H}(CC_1, PQ)$, where CC_1 is a suitable pair of conjugate points. The harmonic homology with center C and axis the polar c (through C_1) interchanges P and Q while transforming the conic into itself (by Exercise 4 of Section 8.2). Since any tangent through P or Q is transformed into a tangent through Q or P, two mutually accessible points are always of the same type: both on the conic, or both exterior, or both interior. By 8.11, any two points on the conic are mutually accessible. By Exercise 1 of Section 8.1, any two exterior points on a tangent are mutually accessible. Let P and Q be any two exterior points whose join is *not* a tangent, and let either of the tangents from P meet either of the tangents from Q in R. Since R is accessible from both P and Q, Exercise 6 shows that P and Q are accessible from each other. Thus the three answers are:
   (i) all the points on the conic,
   (ii) all the exterior points,
   (iii) a set of interior points (possibly all, as we shall see in Section 12.6, Exercise 9, and Section 12.7, Exercise 7).

8. (i) On a given line through P, there are $(q + 1)/2$ points accessible from P: one for each pair of conjugate points on the line. (Two such pairs could not both yield the same point Q; for then the involution of pairs of conjugate points could be described as $(PP)(QQ)$, whereas we know that it is elliptic.)
   (ii) On each of the $q + 1$ lines through P, there are $(q + 1)/2$ points accessible from P, namely P itself and $(q - 1)/2$ others. Hence the total number of points accessible from P is
$$1 + (q + 1)(q - 1)/2 = (q^2 + 1)/2.$$

9. We use *reductio ad absurdum*. Suppose, if possible, that an elliptic polarity exists. By Exercise 6, accessibility is an "equivalence relation" (G. Birkhoff and S. MacLane, *A Survey of Modern Algebra*, 2nd ed., Macmillan, New York, 1953, pp. 155–156). Therefore the $q^2 + q + 1$ points in $PG(2, q)$ can be distributed into a certain number of classes, each consisting of $(q^2 + 1)/2$ mutually accessible points. Dividing $q^2 + q + 1$ by $(q^2 + 1)/2$ (and remembering that q is an odd number, greater than 1), we obtain the quotient 2 and *a remainder* q, which is absurd.

Section **11.1**

1. X coincides with P.
2. See Exercise 5 of Section 9.2.

Section **11.2**

The range and the axial pencil; the flat pencil.

Section **11.3**

(i) No. (ii) No. (iii) Yes.

Section **11.4**

Yes.

Section **11.5**

If two axial pencils belong to a bundle, they have a common plane.

Section **11.6**

(2.14) If AB and CD are either intersecting or parallel, then AC and BD are either intersecting or parallel.

(2.16) If two planes are parallel, their common points form a line at infinity.

Section **11.7**

The sphere on PQ as diameter.

Section **12.1**

When the triad (x_1, x_2, x_3) is interpreted as a point in affine space, represented by Cartesian coordinates, the equivalent triads all lie on a line through the origin, and the equation $\{Xx\} = 0$ represents a plane through the origin. When we pass from affine space to the projective plane, it is thus natural to use x_1, x_2, x_3 as homogeneous coordinates for a point, and to regard $\{Xx\} = 0$ as the equation for a line.

Section **12.2**

1. $(0, 1, -1)$, $(-1, 0, 1)$, $(1, -1, 0)$.
2. Yes, they all pass through $(1, -1, 1)$.
3. $[1, 1, -2]$; $(0, 2, 1)$.
4. $(0, x_2, x_3)$.
5. $(-1, 1, 1)$, $(1, -1, 1)$, $(1, 1, -1)$.
6. The condition for $(\mu y + z)$ to lie on $[X]$ is $\mu\{Xy\} + \{Xz\} = 0$, or $\mu = -\{Xz\}/\{Xy\}$.
7. $AS = [0, x_3, -x_2], \ \ldots, \ AD = [0, x_3, x_2], \ \ldots,$
   $P = (-x_1, x_2, x_3), \ \ldots, \ D = (0, x_2, -x_3), \ \ldots,$
   $S = (1/x_1, 1/x_2, 1/x_3)$.
8. $F = (x_1, -x_2, 0)$.

Section **12.3**

1. In the notation of Figure 2.5A, we may take

$$A = (y), \quad B = (z), \quad C = (y + z), \quad R = (x), \quad Q = (x + y).$$

Then S, on both BQ and CR, must be $(x + y + z)$; P, on both BR and AS, must be $(x + z)$; and F, on both AB and PQ, must be $(y - z)$.

2. In Exercise 1, write μy for y. The point $(\mu y - z)$ is, of course, the same as $(-\mu y + z)$. The harmonic conjugate of $[\mu Y + Z]$ with respect to $[Y]$ and $[Z]$ is $[-\mu Y + Z]$.

3. This is the correspondence between harmonic conjugates with respect to (y) and (z).

4. Any projectivity on $(y)(z)$ is expressible as $(\mu y + z) \barwedge (\mu y' + z')$, where $(y') = (\alpha y + \gamma z)$, $(z') = (\beta y + \delta z)$ and, to make these points distinct,

$$\alpha\delta - \beta\gamma \neq 0.$$

Since $\mu y' + z' = \mu(\alpha y + \gamma z) + \beta y + \delta z = (\alpha\mu + \beta)y + (\gamma\mu + \delta)z,$

the point $(\mu y' + z')$ is $(\mu' y + z)$, where

$$\mu' = (\alpha\mu + \beta)/(\gamma\mu + \delta).$$

5.
$$\frac{x'_1}{x'_2} = \frac{\alpha x_1 + \beta x_2}{\gamma x_1 + \delta x_2} \qquad (\alpha\delta \neq \beta\gamma)$$

or $\rho x'_1 = \alpha x_1 + \beta x_2$, $\rho x'_2 = \gamma x_1 + \delta x_2 \quad (\rho \neq 0)$.

6. In the notation of Exercise 4, the relation between μ and μ' may be expressed in the form

$$\gamma\mu\mu' - \alpha\mu + \delta\mu' - \beta = 0 \qquad (\alpha\delta \neq \beta\gamma).$$

(i) This is an involution if μ and μ' are interchangeable, that is if $-\alpha = \delta$. In other words, the general involution is given by

$$a\mu\mu' + b(\mu + \mu') + c = 0 \qquad (b^2 \neq ac).$$

(ii) Any invariant point is given by a root of the quadratic equation $\gamma\mu^2 - (\alpha - \delta)\mu - \beta = 0$; therefore the projectivity is parabolic if

$$(\alpha - \delta)^2 + 4\beta\gamma = 0.$$

7. This is simply a restatement of the above "algebraic version of the fundamental theorem." In particular, the cross ratio of a harmonic set is -1.

8. Since $[Y]$ passes through $(y + z)$, and $[Z]$ through $(\mu y + z)$, we have

$$\{Yy\} + \{Yz\} = 0, \qquad \mu\{Zy\} + \{Zz\} = 0,$$

and the given expression reduces at once to μ.

9. Taking (y), (z), $[Y]$, $[Z]$ to be $(a, 1, 0)$, $(b, 1, 0)$, $[1, -c, 0]$, $[1, -d, 0]$, we have

$$\frac{\{yY\}\{zZ\}}{\{yZ\}\{zY\}} = \frac{(a-c)(b-d)}{(a-d)(b-c)}.$$

Section **12.4**

1. (i) $x'_1 = px_1$, $x'_2 = qx_2$, $x'_3 = rx_3$.
   (ii) $x'_1 = -x_1 + x_2 + x_3$, $x'_2 = x_1 - x_2 + x_3$, $x'_3 = x_1 + x_2 - x_3$.
   (iii) $x'_1 = x_3$, $x'_2 = x_1$, $x'_3 = x_2$.
   (iv) $x' = x_3$, $x'_2 = -x_1 + x_3$, $x'_3 = -x_2 + x_3$.

2. $$\rho x_i = \sum c_{ij} x'_j, \qquad \sigma X'_j = \sum c_{ij} X_i$$

   or, solving the former set of equations in terms of C_{ij}, the cofactor of c_{ij} in the determinant 12.44,

   $$\rho' x'_j = \sum C_{ij} x_i.$$

3. Corresponding sides of the two triangles meet in the three points

   $$(0, q-1, 1-r), \qquad (1-p, 0, r-1), \qquad (p-1, 1-q, 0),$$

   which are collinear by the criterion 12.21 or, still more simply, by addition.

4. The lines A_1B_1 and A_3B_2 meet in $C_1 = (1, 1, 1)$. Therefore B_3 is $(1, q, 1)$ for a suitable value of q. The three lines A_3B_1, A_1B_3, A_2B_2, having equations

   $$x_1 = px_2, \qquad x_2 = qx_3, \qquad x_3 = rx_1,$$

   all pass through the same point C_2 if $pqr = 1$. The three lines A_3B_3, A_2B_1, A_1B_2, having equations

   $$x_2 = qx_1, \qquad x_1 = px_3, \qquad x_3 = rx_2,$$

   all pass through the same point C_3 if $qpr = 1$. These two conditions agree, since $pq = qp$. (The connection between Pappus's theorem and the commutative law of multiplication is one of the many important discoveries of David Hilbert, 1862–1943.)

Section **12.5**

1. If the correlation 12.51 transforms $(1, 0, 0)$ into $[1, 0, 0]$, $(0, 1, 0)$ into $[0, 1, 0]$, and $(0, 0, 1)$ into $[0, 0, 1]$, we must have $c_{ij} = 0$ whenever $i \neq j$. Therefore $c_{ij} = c_{ji}$, and the correlation is a polarity.

2. Setting $[X] = [Y] = [0, 0, 1]$ in 12.54, we see that the condition for this line to be self-conjugate is $C_{33} = 0$, or

   $$c_{11}c_{22} - c_{12}^2 = 0.$$

Thus the hypothesis of 7.13 requires $c_{11}c_{22} - c_{12}^2 \neq 0$. Setting $(x) = (\mu, 1, 0)$ and $(y) = (\mu', 1, 0)$ in 12.53, we see that the condition for these two points to be conjugate is

$$c_{11}\mu\mu' + c_{12}(\mu + \mu') + c_{22} = 0.$$

By Exercise 6(i) of Section 12.3, this relation between μ and μ' represents an involution provided $c_{12}^2 \neq c_{11}c_{22}$.

3. Setting $\mu = 1$ and $\mu' = -1$ in the above solution of Exercise 2, we see that the condition for the opposite vertices $(\pm 1, 1, 0)$ to be conjugate is

$$c_{11} = c_{22}.$$

Similarly, if $(0, \pm 1, 1)$ are conjugate, $c_{22} = c_{33}$. Hence $c_{11} = c_{33}$, and this makes $(\pm 1, 0, 1)$ conjugate.

4. The polars $[c_{11}, c_{21}, c_{31}]$, $[c_{12}, c_{22}, c_{32}]$, $[c_{13}, c_{23}, c_{33}]$ of the vertices meet the respectively opposite sides $[1, 0, 0]$, $[0, 1, 0]$, $[0, 0, 1]$ in the collinear points

$$(0, c_{31}, -c_{12}), \qquad (-c_{23}, 0, c_{12}), \qquad (c_{23}, -c_{31}, 0).$$

5. In vector notation, $(1, 1, 1) + (-1, 1, 1) = 2(0, 1, 1)$, and so on. Since the sides of the two triangles are

$$[-1, 1, 1]\,[1, -1, 1]\,[1, 1, -1] \quad \text{and} \quad [0, 1, 1]\,[1, 0, 1]\,[1, 1, 0],$$

they are polar triangles for the "natural" polarity $X_i = x_i$ which transforms each point (or line) into the line (or point) that has the same coordinates. This is the algebraic counterpart of Section 7.8. The remaining points and lines of the Desargues configuration are $(0, 1, -1)$, $(-1, 0, 1)$, $(1, -1, 0)$, and $[0, 1, -1]$, $[-1, 0, 1]$, $[1, -1, 0]$. (Reference **8b**, p. 106.)

6. The point (y) is accessible from (z) if and only if there exists a number μ such that the harmonic conjugates $(y \pm \mu z)$ are conjugate for the polarity. Replacing x and y in the expression (xy) by $y + \mu z$ and $y - \mu z$, we see that the desired condition for conjugacy is $(yy) - \mu(yz) + \mu(zy) - \mu^2(zz) = 0$, that is,

$$(yy) = \mu^2(zz)$$

(cf. Reference **7**, pp. 37–38, 192).

Section **12.6**

1. When $(xx) = 2(x_2x_3 + x_3x_1 + x_1x_2)$, the condition for conjugacy $(xy) = 0$ becomes $(x_2y_3 + y_2x_3) + (x_3y_1 + y_3x_1) + (x_1y_2 + y_1x_2) = 0$. Setting $x_3 = y_3 = 0$, we see that, on the secant $[0, 0, 1]$, $(x_1, x_2, 0)$ is conjugate to $(x_1, -x_2, 0)$ (see Exercise 8 of Section 12.2).

2. The tangent at (1, 0, 0) to the conic

$$c_{22}x_2^2 + 2c_{23}x_2x_3 + 2c_{31}x_3x_1 + 2c_{12}x_1x_2 = 0$$

is $c_{31}x_3 + c_{12}x_2 = 0$, which coincides with $x_3 = 0$ if $c_{12} = 0$. Similarly, the tangent at (0, 0, 1) is $x_1 = 0$ if $c_{23} = 0$. Finally, the conic

$$c_{22}x_2^2 + 2c_{31}x_3x_1 = 0$$

passes through (1, 1, 1) if $c_{22} = -2c_{31}$, and we are left with the equation

$$x_2^2 = x_3x_1.$$

3. The point of intersection $(1, -\mu, \mu^2)$ traces the conic $x_2^2 = x_3x_1$.

4. (i) $X_2^2 + 2X_3X_1 = 0$. (ii) $X_2^2 = 4X_3X_1$.

5. Let PQR be the triangle of reference, and O the unit point (1, 1, 1). The point (x_1, x_2, x_3) whose locus we seek is the trilinear pole of the line $[x_1^{-1}, x_2^{-1}, x_3^{-1}]$, which passes through O if

$$x_1^{-1} + x_2^{-1} + x_3^{-1} = 0.$$

Hence the locus is the conic 12.62.

6. If $(x)(y)$ is a tangent, let $(\mu x + y)$ be its point of contact. Since this is conjugate to both (x) and (y), we have

$$\mu(xx) + (xy) = 0, \qquad \mu(xy) + (yy) = 0.$$

Eliminating μ, we obtain the locus

$$(xx)(yy) - (xy)^2 = 0,$$

which is the combined equation of the two tangents that pass through (y). Dually, if $[Y]$ is a secant PQ of the conic $[XX] = 0$,

$$[XX][YY] - [XY]^2 = 0$$

is the combined equation of the two points P and Q.

7. Let (y) and (z) be the two exterior points, so that the points of contact are the intersections of the original conic $(xx) = 0$ with the polars $(xy) = 0$ and $(xz) = 0$. The conic

$$(xx)(yz) - (xy)(xz) = 0$$

evidently passes through all these six points. Dually, two secants $[Y]$, $[Z]$, and the tangents at their four "ends," all touch the conic

$$[XX][YZ] - [XY][XZ] = 0.$$

8. By Exercise 7(ii) at the end of Chapter 10, a point (y) is exterior if and only if it is accessible from (z). By Exercise 6 of Section 12.5, this condition means that (yy), like (zz), is a nonzero square.

   The conic 12.61 has tangents $x_1 = x_3$ and $x_2 = x_3$, meeting in the point $(1, 1, 1)$, for which $(zz) = 1$. Therefore (y) is exterior or interior according as $y_1^2 + y_2^2 - y_3^2$ is a nonzero square or a nonsquare.

9. In the field of real numbers, the product of any two negative numbers is positive. But in the field of rational numbers, the product of two nonsquares is not necessarily a square. To obtain an infinity of mutually inaccessible points, we can take $(xx) = x_1^2 + x_2^2 - x_3^2$ and choose points (y) for which the values of (yy) run through the sequence of primes.

Section **12.7**

1. (0, 0, 1), (0, 1, 0), (0, 1, 1), (0, 1, 4), (0, 2, 1), (0, 1, 2).

2. [4, 0, 1], [1, 4, 0], [1, 3, 1], [1, 1, 3], [3, 1, 1], [0, 1, 4].

3. (i) Since this collineation has the effect

$$\begin{array}{ll} (1,0,0) \to (0,\ 1,\ 0), & [1,0,0] \to [\ 0,1,\ 2], \\ (0,1,0) \to (1,-2,\ 1), & [0,1,0] \to [\ 1,0,\ 0], \\ (0,0,1) \to (0,\ 1,\ 2), & [0,0,1] \to [\ 2,0,-2], \\ (1,1,1) \to (1,\ 0,-2), & [1,1,1] \to [-2,1,\ 0], \end{array}$$

its equations are

$$\begin{array}{ll} x'_1 = x_2, & X'_1 = X_2 + 2X_3, \\ x'_2 = x_1 - 2x_2 + x_3, & X'_2 = X_1, \\ x'_3 = x_2 + 2x_3, & X'_3 = 2X_1 - 2X_3. \end{array}$$

(Remember that, in this finite arithmetic, $-1 = 4$.)

(ii) In this case

$$\begin{array}{ll} (1,0,0) \to (-1,-1,-1), & [1,0,0] \to [-2,1,\ 2], \\ (0,1,0) \to (\ 1,\ 0,\ 1), & [0,1,0] \to [-1,1,\ 0], \\ (0,0,1) \to (\ 2,\ 2,\ 1), & [0,0,1] \to [-1,0,\ 1], \\ (1,1,1) \to (\ 2,\ 1,\ 1), & [1,1,1] \to [\ 1,2,-2]; \end{array}$$

$$\begin{array}{ll} x'_1 = -x_1 + x_2 + 2x_3, & X'_1 = -2X_1 - X_2 - X_3, \\ x'_2 = -x_1 + 2x_3, & X'_2 = X_1 + X_2, \\ x_3' = -x_1 + x_2 + x_3, & X'_3 = 2X_1 + X_3. \end{array}$$

4. $X_1 = x_1, \quad X_2 = x_2, \quad X_3 = x_3.$

5. (i) (1, 0, 2), (0, 2, 1), (2, 1, 0), (2, 0, 1), (0, 1, 2), (1, 2, 0).
   (ii) (0, 1, 1), (0, 1, 4), (4, 0, 1), (2, 1, 0), (1, 2, 0), (1, 0, 1).
   (iii) (0, 0, 1), (1, 0, 0), (1, 1, 1), (2, 1, 3), (3, 1, 2), (1, 4, 1).
   (iv) (0, 0, 1), (0, 1, 0), (1, 0, 0), (1, 2, 1), (2, 1, 1), (1, 1, 2).
   (v) (0, 1, 1), (1, 1, 0), (4, 1, 1), (1, 1, 4), (1, 4, 1), (1, 0, 1).

6. The field must admit an element whose square is -3; for instance, such a configuration exists in the complex plane but not in the real plane, in $PG(2, p^k)$ if and only if p is a prime congruent to 1 or 3 modulo 6. Following the hint, we obtain $F = CD \cdot EH$, $G = BH \cdot DE$. The collinearity of FGA makes

$$\omega^2 + \omega + 1 = 0.$$

The joins of "opposite" points all pass through $(-\omega, 1, 1)$. In the case of $PG(2, 3)$, we have $\omega = 1$, and the $(9_4, 12_3)$ uses up all the points except

$$(0, 1, 2), \quad (2, 0, 1), \quad (1, 2, 0) \quad \text{and} \quad (1, 1, 1),$$

which are the points on the remaining line [1, 1, 1].

7. We normalize the equation of the conic, as in Exercise 8 of Section 12.6, and then apply Exercise 6 of Section 12.5, recalling that, in a finite field, the product of any two nonsquares is a square. In fact, since any nonzero square a^2 has two distinct square roots $\pm a$, the number of nonzero squares is just half the number of nonzero elements. Thus the $q - 1$ nonzero elements of the finite field consist of just $(q - 1)/2$ squares, say $s_1, s_2, \ldots,$ and the same number of nonsquares, say $n_1, n_2, \ldots.$ Multiplying all these $q - 1$ elements by any one of them, we obtain the same $q - 1$ elements again (usually in a different arrangement). Since the $(q - 1)/2$ elements $s_1n_k, s_2n_k, \ldots$ (for a given nonsquare n_k) are the n's, the remaining $(q - 1)/2$ elements $n_1n_k, n_2n_k, \ldots$ must be the s's. (Reference **10a**, p. 70.)

Section **12.8**

1. (i) $x'_1 = -x_1, \quad x'_2 = -x_2$ (as in 12.81 with $\mu = -1$).
   (ii) $x'_1 = 2a_1 - x_1, \quad x'_2 = 2a_2 - x_2.$

2. A rotation about the origin.

3. (i) $x'_1 = x_1, \quad x'_2 = -x_2.$
   (ii) $x'_1 = x_2, \quad x'_2 = x_1.$

References

1. H. F. Baker, *An Introduction to Plane Geometry*, Cambridge University Press, 1943.
2. W. W. R. Ball and H. S. M. Coxeter, *Mathematical Recreations and Essays* (12th ed.), University of Toronto Press, in press.
3. E. T. Bell, *Men of Mathematics*, Simon and Schuster, New York, 1937.
4. J. L. Coolidge, *A History of the Conic Sections and Quadric Surfaces*, Clarendon Press, Oxford, 1945.
5. ——— *The Mathematics of Great Amateurs*, Clarendon Press, Oxford, 1949.
6. Richard Courant and Herbert Robbins, *What is Mathematics?*, Oxford University Press, New York, 1958.
7. H. S. M. Coxeter, *The Real Projective Plane* (2nd ed.), Cambridge University Press, 1955.
8. ——— *Introduction to Geometry*, Wiley, New York, 1969.

8a. H. S. M. Coxeter and S. L. Greitzer, *Geometry Revisited*, Random House, New York, 1967.

8b. H. L. Dorwart, *The Geometry of Incidence*, Prentice-Hall, Englewood Cliffs, N.J., 1966.

9. H. G. Forder, *The Foundations of Euclidean Geometry*, Cambridge University Press, 1927; Dover Publications, New York, 1958.
10. ——— *Geometry* (2nd ed.), Hutchinson's University Library, London, 1960; Harper Torchbooks, London, 1963.

10a. G. H. Hardy and E. M. Wright, *An Introduction to the Theory of Numbers* (4th ed.), Oxford, 1960.

11. Sir Thomas Heath, *A History of Greek Mathematics* (2 vols.), Clarendon Press, Oxford, 1921.

12. D. N. Lehmer, *An Elementary Course in Synthetic Projective Geometry*, Ginn and Company, Boston, 1917.

12a. F. W. Levi, *Geometrische Konfigurationen*, Hirzel, Leipzig, 1929.

13. E. H. Lockwood, *A Book of Curves*, Cambridge University Press, 1961.

14. G. B. Mathews, *Projective Geometry*, Longmans, Green and Company, London, 1914.

15. Beniamino Segre, *Lectures on Modern Geometry*, Cremonese, Rome, 1961.

16. A. S. Smogorzhevskiĭ, *The Ruler in Geometrical Constructions*, Blaisdell Publishing Company, New York, 1962.

17. D. J. Struik, *Lectures on Analytic and Projective Geometry*, Addison-Wesley, Cambridge, Massachusetts, 1953.

18. J. L. Synge, *Science: Sense and Nonsense*, Jonathan Cape, London, 1951.

19. Oswald Veblen and J. W. Young, *Projective Geometry*, Vol. I, Blaisdell Publishing Company, New York, 1966.

20. B. L. van der Waerden, *Einführung in die algebraische Geometrie*, Dover Publications, New York, 1945.

21. ——— *Science Awakening*, Oxford University Press, New York, 1961.

22. A. N. Whitehead, *The Aims of Education and Other Essays*, Williams and Norgate, London, 1929.

23. I. M. Yaglom, *Geometric Transformations*, Vol. III, Random House, New York, 1973.

Index

[References are to pages; principal references are in boldface.]

BY THE SAME AUTHOR

Introduction to Geometry
Wiley, New York

The Real Projective Plane
Cambridge University Press

Non-Euclidean Geometry
University of Toronto Press

Twelve Geometric Essays
Southern Illinois University Press

Regular Polytopes
Dover, New York

Regular Complex Polytopes
Cambridge University Press

(with P. Du Val, H. T. Flather, and J. F. Petrie)
The Fifty-Nine Icosahedra
University of Toronto Press

(with W. W. Rouse Ball)
Mathematical Recreations and Essays
University of Toronto Press

(with S. L. Greitzer)
Geometry Revisited
Random House, New York

(with W. O. J. Moser)
Generators and Relations for Discrete Groups
Springer, Berlin